U.S. CHEMICAL SAFETY AND HAZARD INVESTIGATION BOARD

INVESTIGATION REPORT

REFINERY FIRE INCIDENT

(4 Dead, 1 Critically Injured)

TOSCO AVON REFINERY
MARTINEZ, CALIFORNIA
FEBRUARY 23, 1999

KEY ISSUES:

■ CONTROL OF HAZARDOUS NONROUTINE
MAINTENANCE

■ MANAGEMENT OVERSIGHT AND ACCOUNTABILITY

■ MANAGEMENT OF CHANGE

■ CORROSION CONTROL

REPORT NO. 99-014-I-CA
ISSUE DATE: MARCH 2001

U.S. CHEMICAL SAFETY AND HAZARD INVESTIGATION BOARD
Office of Investigations and Safety Programs
2175 K Street, NW, Suite 400
Washington, DC 20037
(202) 261-7600

U.S. CHEMICAL SAFETY AND HAZARD INVESTIGATION BOARD

INVESTIGATION REPORT

REFINERY FIRE INCIDENT

(4 Dead, 1 Critically Injured)

TOSCO AVON REFINERY

MARTINEZ, CALIFORNIA
FEBRUARY 23, 1999

KEY ISSUES

CONTROL OF HAZARDOUS NONROUTINE MAINTENANCE

MANAGEMENT OVERSIGHT AND ACCOUNTABILITY

MANAGEMENT OF CHANGE

CORROSION CONTROL

REPORT NO. 99-014-I-CA
ISSUE DATE: MARCH 2001

Abstract

This investigation report examines the refinery fire incident that occurred on February 23, 1999, in the crude unit at the Tosco Corporation Avon refinery in Martinez, California. Four workers were killed, and one was critically injured. This report identifies the root and contributing causes of the incident and makes recommendations for control of hazardous nonroutine maintenance, management oversight and accountability, management of change, and corrosion control.

The U.S. Chemical Safety and Hazard Investigation Board (CSB) is an independent Federal agency whose mission is to ensure the safety of workers and the public by preventing or minimizing the effects of chemical incidents. CSB is a scientific investigative organization; it is not an enforcement or regulatory body. Established by the Clean Air Act Amendments of 1990, CSB is responsible for determining the root and contributing causes of accidents, issuing safety recommendations, studying chemical safety issues, and evaluating the effectiveness of other government agencies involved in chemical safety. No part of the conclusions, findings, or recommendations of CSB relating to any chemical incident may be admitted as evidence or used in any action or suit for damages arising out of any matter mentioned in an investigation report (see 42 U.S.C. § 7412(r)(6)(G)). CSB makes public its actions and decisions through investigation reports, summary reports, safety bulletins, safety recommendations, special technical publications, and statistical reviews. More information about CSB may be found on the World Wide Web at http://www.chemsafety.gov.

Salus Populi Est Lex Suprema
People's Safety is the Highest Law

Information about available publications may be obtained by contacting:
U.S. Chemical Safety and Hazard
Investigation Board
Office of Congressional
and Public Affairs
2175 K Street, NW, Suite 400
Washington, DC 20037
(202) 261-7600

CSB publications may be purchased from:
National Technical Information
Service
5285 Port Royal Road
Springfield, VA 22161
(800) 553-NTIS or
(703) 487-4600
Email: info@ntis.fedworld.gov

For international orders, see:
http://www.ntis.gov/support/
cooperat.htm.

For this report, refer to NTIS
number PB2001-104050.

Contents

Contents (cont'd)

Figures

Acronyms and Abbreviations

AIChE	American Institute of Chemical Engineers
API	American Petroleum Institute
Cal/OSHA	California Department of Industrial Relations, Division of Occupational Safety and Health
CCPS	Center for Chemical Process Safety
CFR	Code of Federal Regulations
CSB	U.S. Chemical Safety and Hazard Investigation Board
EPA	U.S. Environmental Protection Agency
°F	Degrees Fahrenheit
HAZOP	Hazard and operability
HSE	Health and Safety Executive (United Kingdom)
ICE	Institution of Chemical Engineers (United Kingdom)
LCV	Level control valve
MOC	Management of change
MSDS	Material Safety Data Sheet
NFPA	National Fire Protection Association
NPRA	National Petrochemical & Refiners Association
NTIS	National Technical Information Service
NTSB	National Transportation Safety Board
OSHA	Occupational Safety and Health Administration
PACE	Paper, Allied-Industrial, Chemical & Energy Workers International Union
psig	Pounds per square inch gage
PPE	Personal protective equipment
PSM	Process safety management
SCBA	Self-contained breathing apparatus
UDS	Ultramar Diamond Shamrock Corporation
U.S.C.	United States Code

Executive Summary

ES.1   Introduction

On February 23, 1999, a fire occurred in the crude unit at Tosco Corporation's Avon oil refinery in Martinez, California. Workers were attempting to replace piping attached to a 150-foot-tall fractionator[1] tower while the process unit was in operation. During removal of the piping, naphtha[2] was released onto the hot fractionator and ignited. The flames engulfed five workers located at different heights on the tower. Four men were killed, and one sustained serious injuries.

Ultramar Diamond Shamrock Corporation (UDS) purchased the facility in September 2000 and renamed it the Golden Eagle refinery.

Because of the serious nature of this incident, and the fact that another fatality had occurred at the Avon facility in 1997, the U.S. Chemical Safety and Hazard Investigation Board (CSB) initiated an investigation to determine the root and contributing causes of the incident and to issue recommendations to help prevent similar occurrences.

ES.2   Incident

On February 10, 1999, a pinhole leak was discovered in the crude unit on the inside of the top elbow of the naphtha piping, near where it was attached to the fractionator at 112 feet above grade.[3] Tosco personnel responded immediately, closing four valves in an attempt to isolate the piping. The unit remained in operation.

Subsequent inspection of the naphtha piping showed that it was extensively thinned and corroded. A decision was made to replace a large section of the naphtha line.[4] Over the 13 days between the discovery of the leak and the fire, workers made numerous

---

[1] A fractionator is an oil refinery processing vessel that separates preheated hydrocarbon mixtures into various components based on boiling point. The separated components are referred to as fractions or cuts. Inside the fractionator, some trays draw off the fractions as liquid hydrocarbon products (such as naphtha), and piping transports them to storage or for further processing.

[2] Petroleum naphtha is a highly flammable mixture of liquid hydrocarbons drawn off as a cut from the fractionator tower.

[3] "Above grade" refers to the vertical distance from ground level at the point upon which equipment rests.

[4] The term "naphtha line" is synonymous with naphtha piping. "Naphtha draw line" was also used at the facility to refer to the naphtha piping. The draw line takes or "draws" naphtha product from the 38th tray of the fractionator, where it flows through a level control valve to the naphtha stripper vessel.

unsuccessful attempts to isolate and drain the naphtha piping. The pinhole leak reoccurred three times, and the isolation valves were retightened in unsuccessful efforts to isolate the piping. Nonetheless, Tosco supervisors proceeded with scheduling the line replacement while the unit was in operation.

On the day of the incident, the piping contained approximately 90 gallons of naphtha, which was being pressurized from the running process unit through a leaking isolation valve. A work permit authorized maintenance employees to drain and remove the piping. After several unsuccessful attempts to drain the line, a Tosco maintenance supervisor directed workers to make two cuts into the piping using a pneumatic saw.[5] After a second cut began to leak naphtha, the supervisor directed the workers to open a flange[6] to drain the line. As the line was being drained, naphtha was suddenly released from the open end of the piping that had been cut first. The naphtha ignited, most likely from contacting the nearby hot surfaces of the fractionator, and quickly engulfed the tower structure and personnel.

ES.3   Key Findings

The hot process unit provided multiple sources of ignition, some as close as 3 feet from the pipe removal work.

1. The removal of the naphtha piping with the process unit in operation involved significant hazards. This nonroutine[7] work required removing 100 feet of 6-inch pipe containing naphtha, a highly flammable liquid. Workers conducting the removal were positioned as high as 112 feet above ground, with limited means of escape. The hot process unit provided multiple sources of ignition, some as close as 3 feet from the pipe removal work. One isolation valve could not be fully closed, which indicated possible plugging.

    On three occasions prior to the incident, the naphtha pipe resumed leaking from the original pinhole and felt warm to the

---

[5] A pneumatic saw is a cutting device that is energized by air pressure rather than electrical energy.

[6] A flange is a rim on the end of a section of piping or equipment used for attachment to other piping and equipment.

[7] The Center for Chemical Process Safety (CCPS) defines "nonroutine work" as unscheduled maintenance work that necessitates immediate repair and may introduce additional hazards (CCPS, 1995b; p. 212). One example is "breakdown maintenance," where equipment is operated until it fails. In this incident, the February 10 naphtha draw line leak is an example of breakdown maintenance.

touch, indicating that one or more isolation valves were leaking. Numerous attempts to drain the piping were unsuccessful; a failed attempt to ream out the drain lines and the removal of a small section of pipe confirmed that the line was extensively plugged. On seven occasions, the downstream naphtha stripper vessel filled–indicating probable isolation valve leakage.

2. The naphtha pipe that was cut open during the repair work was known by workers and the maintenance supervisor to contain flammable liquid. Although Tosco procedures required piping to be drained, depressured, and flushed prior to opening,[8] this was not accomplished because extensive plugging prevented removal of the naphtha. The procedures did not specify an alternative course of action if safety preconditions, such as draining, could not be met. Although the hot process equipment was close to the removal work, Tosco's procedures and safe work permit did not identify ignition sources as a potential hazard. The permit also failed to identify the presence of hazardous amounts of benzene in the naphtha.

3. The naphtha stripper vessel level control bypass valve was leaking, which prevented isolation of the line from the operating process unit. As a result, the running unit pressurized the naphtha piping. Excessive levels of corrosive material and water in the naphtha line and operation of the bypass valve in the partially open position for prolonged periods led to erosion/corrosion of the valve seat and disk. Excessive levels of corrosives and water also produced plugging in the piping and led to the initial leak.

4. Tosco's job planning procedures did not require a formal evaluation of the hazards of replacing the naphtha piping. The pipe repair work was classified as low risk maintenance. Despite serious hazards caused by the inability to drain and isolate the line–known to supervisors and workers during the week prior to the incident–the low risk classification was not reevaluated, nor did management formulate a plan to control the known hazards.

5. Tosco's permit for the hazardous nonroutine work was authorized solely by a unit operator on the day of the incident. Operations

Despite serious hazards caused by the inability to drain and isolate the line–known to supervisors and workers during the week prior to the incident–the low risk classification was not reevaluated, nor did management formulate a plan to control the known hazards.

---

[8] Tosco Avon Safety Procedure S-5, Safety Orders, Departmental Safe Work Permits, October 19, 1998.

supervisors were not involved in inspecting the job site or reviewing the permit.

6. Operations supervisors and refinery safety personnel were seldom present in the unit to oversee work activities. On the morning of the incident, prior to the fire, one operations supervisor briefly visited the unit, but he did not oversee the work in progress and no safety personnel visited the unit. The maintenance supervisor was the only management representative present during the piping removal work.

The U.S. Environmental Protection Agency (EPA) similarly determined that a lack of operations supervisory oversight during safety critical activities was one of the causes of a previous Avon refinery incident, a 1997 explosion and fire at the hydrocracker, which resulted in one fatality (USEPA, 1998; pp. viii, 65).[9]

7. In the 3 years prior to the incident, neither Tosco's corporate safety group nor Avon facility management conducted documented audits of the refinery's line breaking,[10] lockout/tagout,[11] or blinding[12] procedures and practices.

8. Tosco did not perform a management of change (MOC)[13] review to examine potential hazards related to process changes, including operating the crude desalter[14] beyond its design parameters, excessive water in the crude feedstock,[15] and

In the 3 years prior to the incident, neither Tosco's corporate safety group nor Avon facility management conducted documented audits of the refinery's line breaking, lockout/tagout, or blinding procedures and practices.

---

[9] The EPA report states: "Supervision was not present at the unit even though there had been a succession of operating problems just prior to the final temperature excursion that led to the explosion and fire."

[10] "Line breaking" refers to equipment opening.

[11] "Lockout/tagout" refers to a program to control hazardous energy during the servicing and maintenance of machinery and equipment. Lockout refers to the placement of a locking mechanism on an energy-isolating device, such as a valve, so that the equipment cannot be operated until the mechanism is removed. Tagout refers to the secure placement of a tag on an energy-isolating device to indicate that the equipment cannot be operated until the tag is removed.

[12] A blind is a piping component consisting of a solid metal plate inserted to secure isolation.

[13] Management of change is a systematic method for reviewing the safety implications of modifications to process facilities, process material, organizations, and standard operating practices.

[14] The desalter vessel removes inorganic salts, water, and suspended solids to reduce corrosion, plugging, and fouling of piping and equipment.

[15] Feedstock is material of varying constituents that is processed in a refinery.

prolonged operation of the bypass valve in the partially open position. Tosco memos and incident reports revealed that management recognized these operational problems and the increased rate of corrosion. However, corrective actions were not implemented in time to prevent plugging and excessive corrosion in the naphtha piping.

ES.4 Root Causes

1. **Tosco Avon refinery's maintenance management system did not recognize or control serious hazards posed by performing nonroutine repair work while the crude processing unit remained in operation.**

   - Tosco Avon management did not recognize the hazards presented by sources of ignition, valve leakage, line plugging, and inability to drain the naphtha piping. Management did not conduct a hazard evaluation[16] of the piping repair during the job planning stage. This allowed the execution of the job without proper control of hazards.

   - Management did not have a planning and authorization process to ensure that the job received appropriate management and safety personnel review and approval. The involvement of a multidisciplinary team in job planning and execution, along with the participation of higher level management, would have likely ensured that the process unit was shut down to safely make repairs once it was known that the naphtha piping could not be drained or isolated.

   - Tosco did not ensure that supervisory and safety personnel maintained a sufficient presence in the unit during the execution of this job. Tosco's reliance on individual workers to detect and stop unsafe work was an ineffective substitute for management oversight of hazardous work activities.

   - Tosco's procedures and work permit program did not require that sources of ignition be controlled prior to opening equipment that might contain flammables, nor did it specify what

The involvement of a multidisciplinary team in job planning and execution, along with the participation of higher level management, would have likely ensured that the process unit was shut down to safely make repairs once it was known that the naphtha piping could not be drained or isolated.

Tosco's reliance on individual workers to detect and stop unsafe work was an ineffective substitute for management oversight of hazardous work activities.

---

[16] A hazard evaluation is a formal analytical tool used to identify and examine potential hazards connected with a process or activity (CCPS, 1992; p. 7).

actions should be taken when safety requirements such as draining could not be accomplished.

2. **Tosco's safety management oversight system did not detect or correct serious deficiencies in the execution of maintenance and review of process changes at its Avon refinery.**

   Neither the parent Tosco Corporation nor the Avon facility management audited the refinery's line breaking, lockout/tagout, or blinding procedures in the 3 years prior to the incident. Periodic audits would have likely detected and corrected the pattern of serious deviations from safe work practices governing repair work and operational changes in process units. These deviations included practices such as:

   - Opening of piping containing flammable liquids prior to draining.

   - Transfer of flammable liquids to open containers.

   - Inconsistent use of blind lists.

   - Lack of supervisory oversight of hazardous work activities.

   - Inconsistent use of MOC reviews for process changes.

ES.5 Contributing Causes

1. **Tosco Avon refinery management did not conduct an MOC review of operational changes that led to excessive corrosion rates in the naphtha piping.**

   Management did not consider the safety implications of process changes prior to their implementation, such as:

   - Running the crude desalter beyond its design parameters.

   - Excessive water in the crude feed.

   - Prolonged operation of the naphtha stripper level control bypass valve in the partially open position.

   These changes led to excessive corrosion rates in the naphtha piping and bypass valve, which prevented isolation and draining of the naphtha pipe.

2. **The crude unit corrosion control program was inadequate.**

   Although Avon refinery management was aware that operational problems would increase corrosion rates in the naphtha line, they did not take timely corrective actions to prevent plugging and excessive corrosion in the piping.

ES.6 Recommendations

Tosco Corporation

Conduct periodic safety audits of your oil refinery facilities in light of the findings of this report. At a minimum, ensure that:

- Audits assess the following:

  ▲ Safe conduct of hazardous nonroutine maintenance

  ▲ Management oversight and accountability for safety

  ▲ Management of change program

  ▲ Corrosion control program.

- Audits are documented in a written report that contains findings and recommendations and is shared with the workforce at the facility.

- Audit recommendations are tracked and implemented.

Ultramar Diamond Shamrock
Golden Eagle Refinery

1. Implement a program to ensure the safe conduct of hazardous nonroutine maintenance. At a minimum, require that:

   - A written hazard evaluation is performed by a multi-disciplinary team and, where feasible, conducted during the job planning process prior to the day of job execution.

   - Work authorizations for jobs with higher levels of hazards receive higher levels of management review, approval, and oversight.

- A written decision-making protocol is used to determine when it is necessary to shut down a process unit to safely conduct repairs.

- Management and safety personnel are present at the job site at a frequency sufficient to ensure the safe conduct of work.

- Procedures and permits identify the specific hazards present and specify a course of action to be taken if safety requirements—such as controlling ignition sources, draining flammables, and verifying isolation—are not met.

- The program is periodically audited, generates written findings and recommendations, and implements corrective actions.

2. Ensure that MOC reviews are conducted for changes in operating conditions, such as altering feedstock composition, increasing process unit throughput, or prolonged diversion of process flow through manual bypass valves.

3. Ensure that your corrosion management program effectively controls corrosion rates prior to the loss of containment or plugging of process equipment, which may affect safety.

American Petroleum Institute (API)
Paper, Allied-Industrial, Chemical & Energy Workers International Union (PACE)
National Petrochemical & Refiners Association (NPRA)

Communicate the findings of this report to your membership.

1.0   Introduction

1.1   Background

On February 23, 1999, a fire occurred in the crude unit at the Tosco Avon oil refinery in Martinez, California. Workers were attempting to replace piping attached to a 150-foot-tall fractionator tower while the process unit was in operation. During removal of the piping, naphtha was released onto the hot fractionator and ignited. The flames engulfed five workers located at different heights on the tower. Four of the workers died, and the fifth was seriously injured. Three of the deceased were contractors–two were employed by a scaffold erection company, and the other worked for a crane company. The fourth fatality and the worker injured were Tosco maintenance employees.

Because of the seriousness of the incident and the fact that there had been a fatal explosion and fire at the refinery 26 months earlier, the U.S. Chemical Safety and Hazard Investigation Board (CSB) launched an investigation to determine the root and contributing causes and to issue recommendations to help prevent similar occurrences.

1.2   Investigative Process

CSB was one of three governmental agencies that investigated the incident. The California Department of Industrial Relations, Division of Occupational Safety and Health (Cal/OSHA), and the Contra Costa County Health Services Department also conducted investigations.[1] The CSB incident investigation team coordinated and shared information with these two agencies.

CSB examined physical evidence at the site, conducted interviews, and reviewed relevant documents (such as a report authored by FTI Anamet (1999), prepared for Cal/OSHA, entitled *Metallurgical Evaluation of Naphtha Draw Line/Valve and Analyses of Petroleum*

---

[1] Through the Department of Industrial Relations, Division of Occupational Safety and Health (Cal/OSHA), California administers its own workplace safety and health program according to provisions of the Federal Occupational Safety and Health Act of 1970 (see 29 CFR 1910). On January 15, 1999, Contra Costa County enacted an Industrial Safety Ordinance to "prevent and reduce the probability of accidental releases of regulated substances that have the potential to cause significant harm to the public health" (Contra Costa County Ordinance Code, Ordinance No. 98-48, Section 1). The ordinance includes a risk management program, a human factors program, and a root cause analysis and incident investigation program. The human factors program was not in effect at the time of the incident. Contra Costa Health Services produced a report on the incident, entitled *Investigation Into the Causes of the Fire of February 23, 1999, at No. 50 Crude Unit, Tosco Avon Refinery*.

Samples From a Crude Unit at the Tosco Avon Refinery). CSB also contracted with The Hendrix Group in Houston, Texas, for assistance with corrosion and mechanical integrity analysis. The American Petroleum Institute (API) and the National Petrochemical & Refiners Association (NPRA) provided good practice information on the safe performance of maintenance work in oil refineries. In response to a CSB request, these two organizations produced a report entitled *Work Authorization in Refineries* (API/NPRA, 2000).

1.3 Tosco Avon Oil Refinery Facility

The Avon refinery is located on a 2,300-acre site near the town of Martinez in Contra Costa County, California. The refinery has been in operation for more than 80 years; its main products are motor fuels such as gasoline and diesel. Tosco Corporation operated the Avon facility from 1976-2000, when it was purchased by Ultramar Diamond Shamrock (UDS) and renamed the Golden Eagle refinery. Tosco is the nation's largest independent refiner of petroleum products and operates seven refineries across the United States.

1.4 Crude Fractionator and Naphtha System

1.4.1 Fractionator

The Avon facility refined crude oil[2] into motor fuels; other products included propane, butane, and fuel oils. The crude unit, or 50 Unit,[3] was originally designed and built in 1946, and had undergone several major capital improvements.

Crude oil fractionation is the initial step in the refining process. It involves splitting crude oil into portions with similar boiling points. The oil is distilled into streams, including natural gasoline, naphtha,

---

[2] Crude oil is a complex mixture of hydrocarbons that varies in composition, quality, and appearance from one producing field to another.

[3] The crude unit where the incident occurred was also referred to as the "50 Unit."

kerosene, diesel, and a heavy oil used as feed for the cracking unit. A series of trays inside the fractionator functions in part to condense the hydrocarbons; in some cases, the trays are used to draw off liquid products from the tower. Processing is continuous. A steady flow of crude oil is pumped into the unit, while the product and feedstock streams are continuously pumped to tanks or other refinery units for further processing.

1.4.2 Naphtha System

At the Avon refinery, naphtha was removed from a tray near the top of the fractionator (112-foot level) into 6-inch steel piping. The naphtha flowed through the piping and a level control valve, and then into the naphtha stripper (Figure 1). From there, it was pumped to storage and the reformer unit for further processing.

In prior years, the stripper had been used to remove lighter hydrocarbons from the naphtha. This practice had been discontinued at the time of the February 23 incident. However, the vapor return line remained in place.

Naphtha Draw From Fractionator

To Naphtha Stripper Vessel

Four-inch bypass globe valve "B"

Drain Valve "F"

Block Valve "C"

Naphtha Stripper Level Control Valve ("D" or LCV-150)

Block Valve "E"

Figure 1. Naphtha stripper level control valve manifold removed to ground level.
The valve at the top right of manifold is the 4-inch bypass valve.

2.0 Description of Incident

The Incident Timeline in Appendix A summarizes the sequence of activities that led to the fire on February 23, 1999.

2.1 Pre-Incident Events

2.1.1 Detection of Naphtha Piping Leak

On February 10, the crude unit was operating routinely when a pinhole leak was detected in the upper section of the naphtha piping. From the ground, the leak was observed to be small and dripping naphtha from the line through the insulation and onto a deck on the fractionator.

2.1.2 Emergency Response and Inspection

Emergency responders decided to attempt to isolate the line to slow or stop the pinhole leak without shutting down the process unit. Operators lowered the pressure in the fractionator and diverted liquid from the naphtha tray. Personnel then donned firefighting "bunker gear"[4] and SCBAs,[5] and closed the block valve (valve A; unless otherwise noted, all valve and flange locations referenced in Section 2.0 are shown in Figure 2). Operators closed the naphtha stripper level control bypass valve (valve B) and two block valves (valves C and E).

Later in the day, the operations supervisor[6] generated an emergency work order. Over the next 13 days, 15 work permits were written for

---

[4] Bunker gear is flame- and heat-resistant clothing.

[5] Self-contained breathing apparatus (SCBA) is respiratory protection worn when the breathable atmosphere may be dangerous to life or health.

[6] At the Avon refinery, the operations supervisor, business team leader, operations shift supervisor, and operations superintendent were four distinct job titles/positions. The operations supervisor worked days and, in particular, was responsible for prioritizing and coordinating maintenance work. He or she reported to the business team leader, who managed an area of the refinery as a business unit, solved day-to-day problems, and implemented long-term projects. The operations shift supervisor (hereinafter referred to as shift supervisor) worked rotating shifts and was the direct supervisor for all operators on his or her crew. The shift supervisor provided both work direction and personnel oversight. Shift supervision and the business team were in separate organizations at the Avon refinery and reported to different operations superintendents.

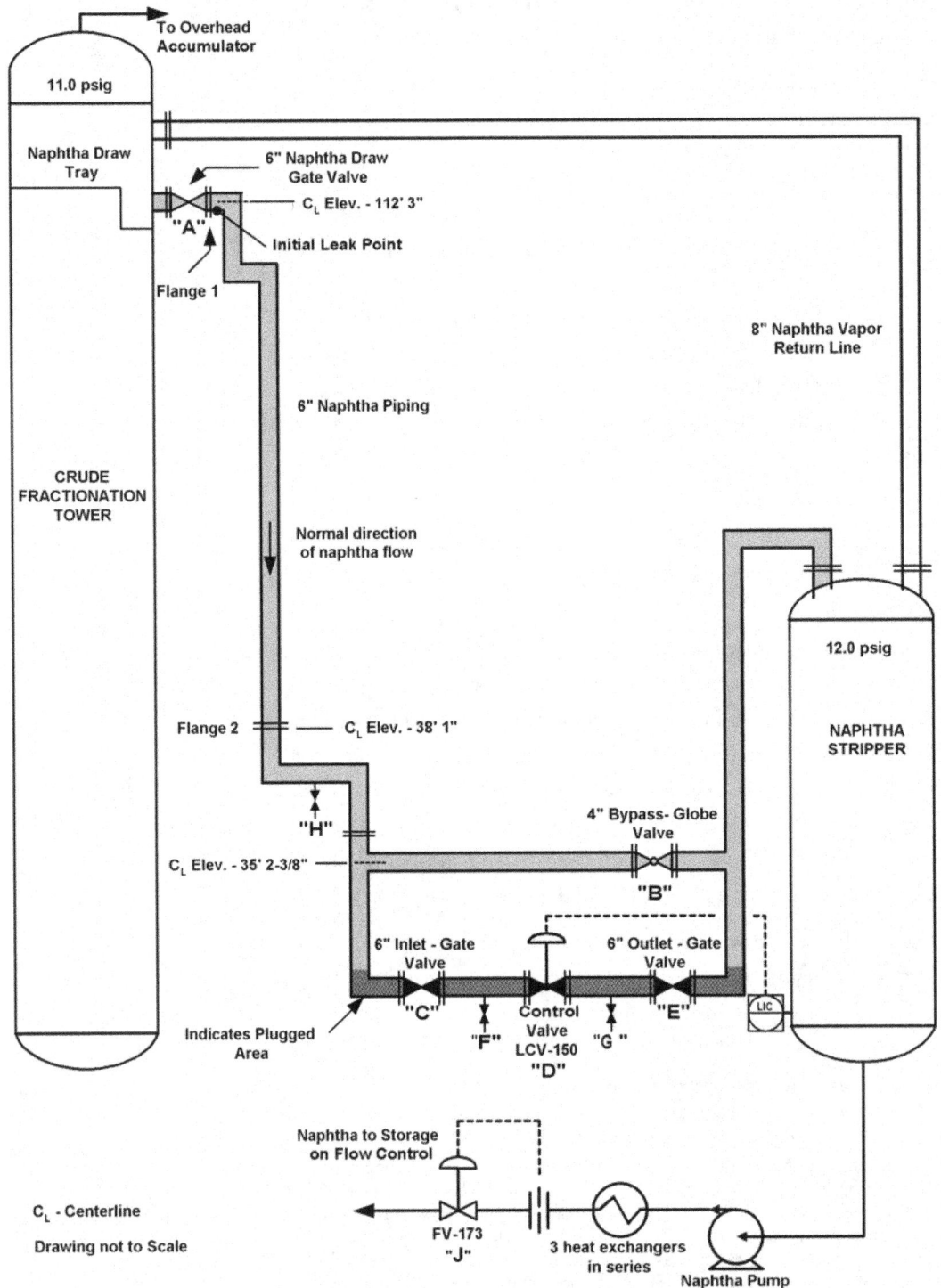

Figure 2. Fractionator and naphtha draw, simplified diagram at time of initial leak.

this job, of which 11 met Tosco's requirements for a special hazard permit (e.g., for inspection radiography, asbestos removal, and lead abatement).

Once the insulation was stripped from the piping, the leak was determined to have originated from a 0.16-inch-long pinhole perforation on the inner radius of an elbow directly downstream of the block valve off the fractionator (valve A), at an elevation of 112 feet. Further inspection using ultrasound and X-ray techniques revealed that much of the piping was severely corroded and thin. Technical staff recommended that the entire line be replaced from valve A to the naphtha stripper.

2.1.3 Recurrence of Leak

On February 13 and 17, operators observed that the leak reoccurred at the original site and that the naphtha piping was warm to the touch. On both occasions, Tosco personnel retightened the piping isolation valves (A and B), and the leak appeared to subside.

On seven occasions from February 10-14, the liquid in the naphtha stripper rose to a high level and was lowered each time by operators opening the naphtha to storage flow control valve (valve J). On the last occasion, they left the valve open to the storage tank to prevent buildup in the stripper; the valve remained open until the day of the fire.

On February 22, while preparing the fractionator area for hot work (i.e., cutting a metal deck to facilitate removal of the piping), the No. 1 operator[7] discovered the naphtha piping again dripping from the original leak point. The piping was hot to the touch. The shift supervisor observed the leak, and a small plug was placed in the hole. After the hot work was finished, the maintenance supervisor directed that the plug be removed.

---

[7] Two or three operators were generally assigned to run the 50 Unit. They worked 12-hour rotating shifts. The No. 1 operator functioned as the lead worker and had the primary responsibility for running the unit safely and according to specifications.

2.1.4 Efforts to Drain and Replace Piping

On February 16 and 17, a No.1 operator attempted to drain the naphtha piping under repair using the drain lines (valves F and G) located on either side of the naphtha stripper level control valve (valve D). The attempt failed; the drain lines appeared to be plugged.

Operators advised the operations supervisor on February 17 that it was not feasible to replace the entire naphtha line from the fractionator to the stripper while the unit was still operating, as recommended by the inspection group. They pointed out that the section of line downstream of block valve E to the stripper could not be isolated because there was no block valve on the naphtha vapor return line.

With input from Tosco inspectors, the operations supervisor determined that the downstream piping did not require immediate replacement. The supervisor considered removal of the line from the fractionator to the control valve (valve D) to be a safe option because of the available isolation valves and drain lines.

On February 18, pipefitters again attempted to clear the drain lines (valves F and G) at the naphtha stripper level control valve by using a reaming device. However, the device broke due to the hardness of the material in the line.

On February 19, the maintenance supervisor directed an operator to issue a permit for removal of the spool piece[8] (from valve D to E) just downstream of the naphtha stripper level control valve. The supervisor was present at the job site during removal of the spool piece. The pipe was plugged solid with a dark, tar-like substance, which also contained large chunks of hard material (Figure 3). A blind flange equipped with a drain valve (valve I; Figure 4) was installed on the downstream side of the control valve, and a solid blind flange[9] was attached on the upstream side of the block valve (valve E).

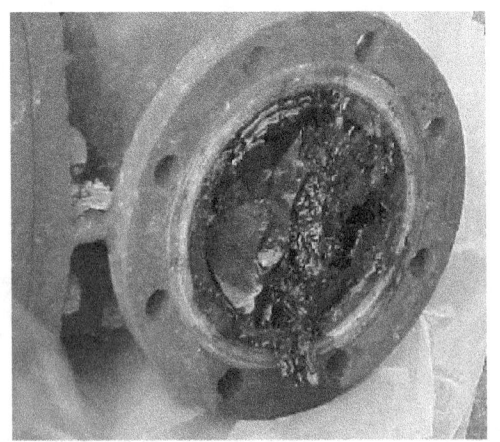

Figure 3. Closeup of material blockage of block valve (C), upstream of LCV-150.

[8] A "spool piece" is a short piece of pipe flanged on both ends to provide for ease of removal or modification.

[9] This flange is a solid plate piping component used for closing an open end of pipe.

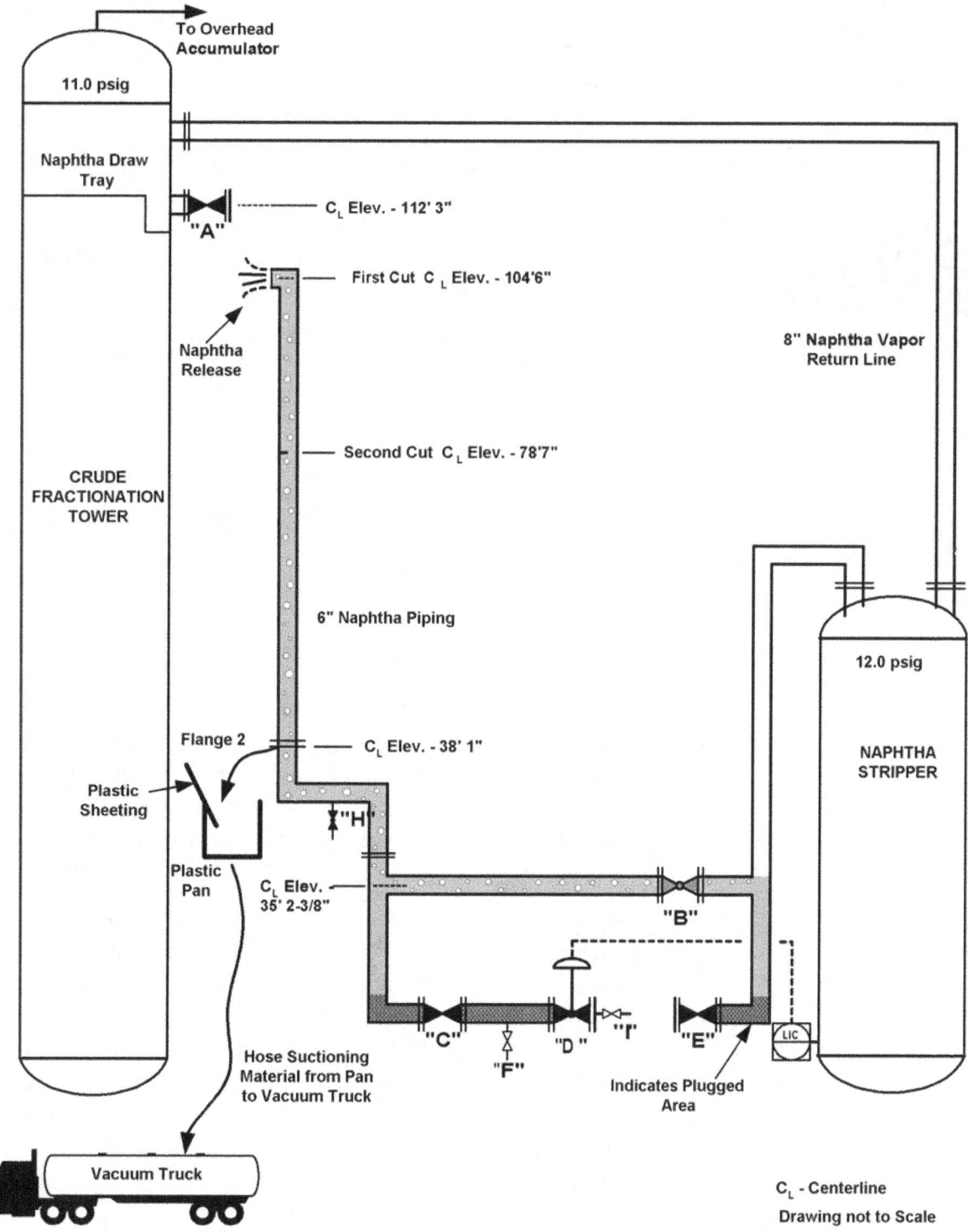

To Overhead Accumulator

11.0 psig

Naphtha Draw Tray

CRUDE FRACTIONATION TOWER

"A"

C_L Elev. - 112' 3"

First Cut C_L Elev. - 104'6"

Naphtha Release

Second Cut C_L Elev. - 78'7"

6" Naphtha Piping

8" Naphtha Vapor Return Line

12.0 psig

NAPHTHA STRIPPER

Flange 2

Plastic Sheeting

Plastic Pan

C_L Elev. - 38' 1"

"H"

C_L Elev. 35' 2-3/8"

"B"

Hose Suctioning Material from Pan to Vacuum Truck

"C"

"F"

"D"

"I"

"E"

LIC

Indicates Plugged Area

Vacuum Truck

C_L - Centerline
Drawing not to Scale

Figure 4. Fractionator and naphtha draw, simplified diagram, draining at lower flange at time of release.

Figure 5. Closeup of stem of block valve (C), upstream of LCV-150, with the valve wheel fully tightened.
The protruding stem shows the valve to be jammed partway open, indicating possible material blockage in the line.

The maintenance supervisor and the workers decided not to remove the spool piece upstream of the control valve (from valve C to D). The supervisor determined that valve C was jammed partway open, and isolation was in doubt (Figure 5). The operator logbook stated that draining would be attempted on Monday, February 22.

However, workers did not attempt to drain the naphtha piping on February 22. The hot work permit for cutting the deck–signed by the shift supervisor, the No. 1 operator, and a maintenance worker–stated that the piping was not drained, locked, or tagged.

The maintenance supervisor and the maintenance lead planner arranged for a vacuum truck to arrive at the job site on February 23 to recover the naphtha.

2.2 The Incident

2.2.1 Job Preparation

On February 23, supervisors, operators, and maintenance workers were aware that the piping contained liquid naphtha. Both the permit readiness sheet and the work permit identified that draining was needed. The No. 1 operator and the maintenance workers inspected the job site and reviewed equipment conditions, and the permit was signed.

In preparation for draining the line, the vacuum truck was placed into position approximately 20 feet from the base of the fractionator. A metal half-barrel was placed under the flange, with the attached drain valve (valve I; Figure 4) downstream of the naphtha stripper level control valve (valve D). A hose was extended from the truck and placed in the barrel. An operator incrementally opened valve D from the control room to assist with draining the line from valve I. Under the direction of their supervisor, the maintenance workers then attempted to open a flange upstream of the control valve. Both efforts to drain the line were unsuccessful.

The maintenance supervisor told the workers present that a section of piping should be cut and removed with the crane. He tapped on the line and stated that he believed the naphtha level was below the proposed cut location. He stated to the operator that listening for differences in the sound at each tap point would identify the liquid level. The operator disagreed and responded that the naphtha should be removed before cutting the pipe.[10]

2.2.2   First Cut and Second Cut

The maintenance supervisor directed workers to unbolt the piping from flange 1, downstream of valve A, and cut a short section of line with a pneumatic saw.[11]   The first cut into the line was 8 feet below

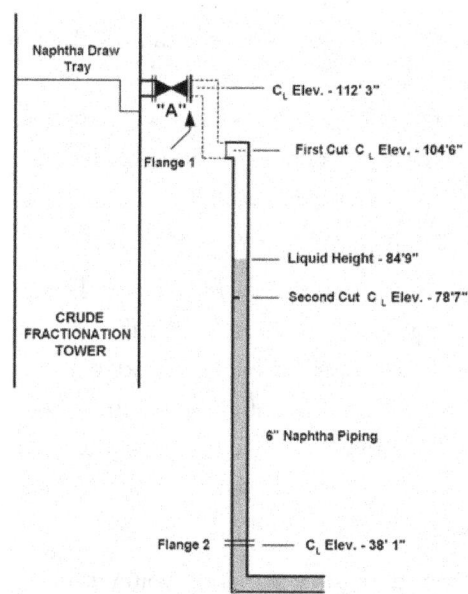

Figure 6. Fractionator and naphtha draw, simplified diagram, first cut and second cut.

[10] Witnesses in the control room in the late morning stated that the No. 1 operator discussed with them his argument with the maintenance supervisor prior to the fire.

[11] The maintenance supervisor, however, denied that he was present or directed the first cut into the naphtha piping.   He stated that he left the unit at 9:00 am.   Other witnesses and the timeline of events contradict this testimony.   For example, the verbal permit log shows the maintenance supervisor signing into the unit at 8:40 am and departing at 9:50 am.   The fact that he directed the work is consistent with his actions both before and after the first cut.   The maintenance supervisor acknowledged that he directed the removal of the spool piece on February 19, and the second cut into the line and the opening of the flanges after lunch on February 23–before the piping had been drained or the isolation verified, contrary to Tosco procedures.

valve A. A blind flange was bolted to valve A. The remaining piping was open at the point of the cut and faced the fractionator (Figure 6).

For the second cut, the maintenance supervisor directed workers to start 26 feet below the location of the first cut (Figure 6). When the saw pierced the inside diameter of the pipe wall, a small amount of liquid began to leak from the line. The worker operating the saw ceased cutting and was sent to obtain a pipe clamp to seal the leak.

2.2.3 Naphtha Release

The maintenance supervisor decided to again attempt to drain the line by opening flange 2, located upstream of the naphtha stripper level control valve (valve D) and within 3 feet of the fractionator (Figure 4). Workers loosened the bolts on flange 2, which allowed liquid to flow. Plastic sheeting was hung to deflect the draining liquid away from the hot fractionator and into an open plastic pan, from which it was suctioned to the vacuum truck.

The personnel conducting the work did not take into account that the naphtha piping was pressurized from the running process unit due to a severe leak through a badly corroded valve (valve B). In the "U"-shaped naphtha piping configuration, the head pressure of the vertical column of liquid functioned as a seal and prevented the process pressure from being released to atmosphere out the open end of the cut pipe.

Once the workers drained a sufficient volume of naphtha from the flange on the vertical run of the piping (flange 2), the pressure from the running process unit leaking through the corroded valve surpassed the reduced head pressure in the line. This resulted in a sudden release of liquid from the open piping at approximately 12:18 pm (Figure 4). The naphtha contacted the hot fractionator and ignited, quickly engulfing the tower structure and personnel.

2.3 Autoignition

The autoignition temperature of a material is defined as the temperature at which its fuel/air mixture will ignite from its own heat source or contact with a hot surface, without spark or flame. Tosco's Material Safety Data Sheets (MSDS) for naphtha listed the autoignition temperature as 450 degrees Fahrenheit (°F). However, the lower half of a crude oil fractionator operates at temperatures of 500 to 650°F, and the noninsulated manways protruding from the Avon refinery fractionator had surface temperatures just slightly below this range.

2.4 Emergency Response

Operators heard the naphtha ignite, used fire monitors to direct a stream of water onto the fire, and began an emergency shutdown of the unit. Within minutes, the Tosco emergency response team was on scene and began firefighting efforts. The Contra Costa Fire and Consolidated Fire Departments responded and were positioned to provide support if requested. The fire burned for about 20 minutes.

Rescue efforts were delayed because of the size of the fire, the risk of re-ignition, and the location of most of the victims on the tower. One worker was pronounced dead at the scene, and the other three victims died at the hospital. The fifth worker jumped away from the flames at an elevated location and sustained serious injuries.

3.0 Analysis of Incident

The conduct of maintenance work in an oil refinery often involves flammable and toxic hazards, which must be carefully controlled to avoid injury to people and the environment (Lees, 1996; p. 21/2). In investigating the Avon refinery incident, CSB found problems with job planning, hazard identification and evaluation, unit shutdown decision making, management oversight, permitting and line breaking, corrosion control and mechanical integrity, and management of change (MOC). CSB used several investigative techniques to analyze the incident, including establishing a timeline (Appendix A) and developing a logic tree diagram (Appendix B).

3.1 Hazardous Nonroutine Maintenance

In process plants, hazardous nonroutine maintenance includes such activities as hot work,[12] hot tap,[13] and work on live flare headers as well as line breaking when isolation and drainage cannot be ensured. The nonmandatory appendix in the Occupational Safety and Health Administration's (OSHA) Process Safety Management (PSM) standard[14] stresses the importance of employers identifying the hazards of nonroutine maintenance in process areas and communicating such hazards to those doing the work.

The 1989 Phillips Houston Chemical Complex fire and explosion, which killed 23 workers, expedited issuance of the PSM standard. Like the 1999 Tosco incident, it involved improper isolation of piping and the failure of a valve during the conduct of hazardous nonroutine maintenance work in a running process unit (OSHA, 1990; pp. iv, ix, 72).

Because nonroutine maintenance is unscheduled, it may present special hazards. One such hazard introduced with breakdown maintenance, such as the job at Tosco, involves limitations on job planning (CCPS, 1995b; p. 212).

---

[12] Hot work is "an operation that can produce a spark or flame or other source of ignition having sufficient energy to cause ignition, where the potential for flammable vapors, gases, or dust exists" (API, 1995b; pp. 2-3).

[13] Hot tapping is "the technique of attaching a mechanical or welded branch fitting to piping or equipment in service, and creating an opening . . . by drilling or cutting a portion of the piping or equipment within the attached fitting" (API, 1995a; p. 1).

[14] Appendix C of 29 CFR 1910.119, Process Safety Management of Highly Hazardous Chemicals.

3.2 Job Planning

In the Avon refinery incident, preparatory maintenance activities such as stripping insulation and inspecting the piping began immediately after the leak appeared to subside. Job planning at the refinery typically involved a site visit and discussions among the maintenance lead planner, maintenance supervisor, and operations supervisor to identify potential problems in advance. However, this planning activity did not occur for the naphtha repair work.

Just an hour after the leak was discovered, a permit was issued to strip insulation from the naphtha piping. The job was initiated without the line being locked or tagged out, depressured, or isolated, as required by Tosco procedures. If a line could not be isolated, the procedures stated that:

> Production, H&S, Inspection and Maintenance representatives must meet and agree on a safe procedure to remove the insulation. If insulation cannot be safely removed while the unit is online, the line or unit must be shut down.[15]

Although inspection, maintenance, and two operations superintendents were present, no meeting was held to discuss control of hazards. Not following insulation removal procedures did not directly cause the fire, but it was indicative of Tosco's practice of not consistently adhering to established maintenance procedures.

Most of the preparatory maintenance work conducted in the 13 days prior to the incident was not listed in the job planning documentation, including the three permitted jobs where workers attempted to drain the piping. There was no mention that the naphtha, as a benzene-containing stream, was a serious health hazard that required specific precautionary measures; nor was it identified that a crane, vacuum truck, and pneumatic saw were to be used in the piping removal.

No job-specific instructions were prepared for the naphtha piping repair work. The job planning documentation lacked necessary information, such as the MSDS for naphtha, a blind list,[16] or the piping and instrumentation diagram showing where blinds were to be inserted. Good practice guidelines for maintenance job planning

---

[15] Tosco Avon Safety Procedure S-36-1, Removing Insulation From Leaking Hydrocarbon Lines, November 20, 1995.

[16] A blind list is a document that specifies the location for blinds to be installed to secure isolation of piping and equipment.

recommend outlining the steps necessary to accomplish the work and identifying the potential hazards of each step (CCPS, 1995b, pp. 249-250; Lees, 1996, p. 21/4).[17]

Adequate planning is also essential for effective isolation of piping and equipment (HSE, 1997; pp. 4, 6, 13).[18] Good practice guidelines emphasize that hazards are most effectively recognized and evaluated in the calm atmosphere of the job planning process rather than during the often stressful environment of job execution (HSE, 1985, p. 11; CCPS, 1995c, p. 17).[19] For example, for hot work—one type of hazardous nonroutine maintenance—API states that the potential hazards should be carefully analyzed as part of pre-job safety planning (API, 1995b; pp. 2-3). Prior to conducting hot tapping, API recommends preparing a written plan that addresses potential hazards and performing the procedure only after careful consideration of alternatives (API, 1995a; pp. 1, 5).

During planning, Tosco management did not effectively identify the serious hazards present in conducting the piping repair in an operating process unit. Despite accumulating evidence of the inability to drain and isolate the piping during the week leading up to the fire, Avon management scheduled the pipe removal for February 23 with the unit in operation and without a plan to control hazards.

Good practice guidelines emphasize that hazards are most effectively recognized and evaluated in the calm atmosphere of the job planning process rather than during the often stressful environment of job execution.

Despite accumulating evidence of the inability to drain and isolate the piping during the week leading up to the fire, Avon management scheduled the pipe removal for February 23 with the unit in operation and without a plan to control hazards.

---

[17] Although this CCPS citation references unit shutdown, the practice is even more appropriate for formal consideration during the safe execution of hazardous nonroutine maintenance in an operating unit.

[18] These good practice guidelines were produced in the United Kingdom on a consensus basis by representatives of industry, government, and labor.

[19] In discussing the management dilemma of production versus process safety, CCPS guidelines state: "The continuity of operations can best be addressed at the planning stage."

3.3 Hazard Identification and Evaluation

3.3.1 Job-Specific Hazards

Significant hazards existed early on in the 13-day naphtha piping repair process:

- The job involved the removal of 100 feet of 6-inch pipe containing naphtha, a highly flammable liquid.[20]

- Approximately 80 feet of the piping ran vertically near the side of the fractionator, whose surface temperature in the lower half of the tower exceeded the autoignition temperature of the naphtha stream to be drained.

- The stem of block valve C, upstream of naphtha stripper level control valve D, protruded approximately 12 threads from the valve wheel when fully tightened, indicating that the valve was partially open and possibly plugged. (Unless otherwise noted, all valve and flange locations referenced in Section 3.0 are shown in Figure 2.)

- The lack of a high point vent downstream of valve A would have made it difficult to purge the naphtha piping. [21]

Tosco classified the naphtha piping repair as low risk, routine maintenance. Management did not recognize or evaluate the inability to isolate, inability to drain, or other hazards in light of conducting the work in an operating process unit.

3.3.1.1 Inability to Isolate

On three occasions prior to the fire, the naphtha piping resumed leaking at the original location and the piping felt warm to the touch, indicating that one or more isolation block valves were leaking. In each instance, the valves were further tightened to try to stop the leak.

---

[20] Tosco's MSDS for the 50 Unit naphtha stated that the National Fire Protection Association (NFPA) flammability hazard rating for the liquid was 4 (on a 0 to 4 scale, with 4 being the most flammable). However, for the 15 work permits authorized for this repair, the NFPA rating either was not provided or was understated as a 2 or 3.

[21] Purging the piping is important not only to remove residual material, but also to reveal possible plugging or solid material in the line, which can trap pressurized or residual liquids and gases (HSE, 1989, pp. 13-14; Kletz, 1989, p. 13).

On February 13, a shift supervisor helped tighten the valves after the leak reoccurred. On February 17, a maintenance supervisor observed the leak reoccur and assisted operators in tightening the isolation valves. That same day, two operators told the operations supervisor and a maintenance supervisor that more than one valve isolating the naphtha piping was potentially leaking. On the morning of February 23, the operations process engineer stated that he suspected that an isolation valve was leaking.

On seven occasions from February 10-14, the naphtha stripper vessel—which was located downstream of the naphtha piping—filled and operators lowered the level. On February 13, a shift supervisor log recorded that the stripper level had been lowered. This log was typically read by other supervisors and was available electronically to other management personnel. The shift supervisor stated that he suspected a valve might have been leaking. Leakage through the isolation valves was the most likely explanation for recurrence of the high level in the naphtha stripper. [22]

3.3.1.2 Inability to Drain

Draining equipment to remove hazardous material and verifying isolation of the line are essential safety requirements prior to maintenance (CCPS, 1995a, p. 310; HSE, 1997, p. 47).[23] From the discovery of the leak to the fire, there were seven failed attempts to drain the naphtha from the piping. Tosco supervisors and workers were aware of the following draining problems:

- On February 16, a No. 1 operator informed the business team leader that the naphtha drain lines were plugged. On February 17, after another unsuccessful draining attempt, two operators discussed plugging in the line with the operations supervisor and

Draining equipment to remove hazardous material and verifying isolation of the line are essential safety requirements prior to maintenance.

---

[22] The naphtha stripper level filling several times in 2-hour intervals, combined with recurrence of the leak, established that the isolation valves were leaking. On February 14, the naphtha flow control valve (valve J) downstream of the stripper was left open, allowing the naphtha to flow out. The stripper remained empty until the day of the fire.
[23] As stated in HSE's *The Safe Isolation of Plants and Equipment:* "Bleeds and vents allow the safe depressurization of parts of the plant when it has been isolated and also enable the integrity of isolations to be checked."

the maintenance supervisor. The operators proposed shutting down the unit to repair the piping. The operator logbook stated that the drain valves were plugged.

- On February 18, the supervisors scheduled maintenance workers to drill out the drain lines. A permit readiness sheet was sent electronically to the operator and the shift supervisor communicating that the drain valves were to be cleared with a reaming device. Permit readiness sheets were also available to management in an electronic bulletin board. After several attempts, maintenance workers informed their supervisor that the material in the piping broke the reaming tool. The operator logbook noted that the attempt to drill out the drain lines was unsuccessful.

- Workers accompanied by the maintenance supervisor removed a small section of piping downstream of the naphtha stripper level control valve (valve D) on February 19. The piping and the drain lines were extensively plugged. Block valve C, upstream of control valve D, was jammed partially open. Both permits executed on February 18 and 19 to drain the piping were checked "job not finished" in the closeout section.

- On February 22, the operations supervisor issued a permit readiness sheet, with input from the maintenance supervisor, stating that draining was part of the work to be performed the following day. A vacuum truck was to be used to collect the naphtha.

- An operations process engineer who visited the unit the morning of the fire was aware that naphtha was still in the piping and was told by operators of the unsuccessful attempts to drain the piping. At the direction of the maintenance supervisor, draining was attempted three times on the morning of February 23.

3.3.1.3 Other Hazards

Another hazard not identified was that the naphtha contained benzene.[24] Because benzene is a carcinogen, Tosco procedures required that equipment be drained to a closed system, away from employees.[25] Maintenance work in the presence of benzene required the use of a special hazard permit with authorization by an operations supervisor. However, Tosco management did not recognize or permit the naphtha piping repair work as a benzene hazard, and these controls were not implemented. Not following these procedures did not directly cause the fire, but demonstrated Tosco's inconsistent adherence to its procedures.

The pipe removal job involved coordinating contractors and workers from different departments, and using a crane in an operating process unit. Furthermore, it required positioning workers where they were potentially subject to the hazard of a sudden release or splashing of flammable liquid. Some workers were located as high as 112 feet above ground. Opening elevated lines is particularly hazardous because of the danger of flammable liquid splashing on personnel or sources of ignition (Kletz, 1989, pp. 14-15; see also CCPS, 1995a, p. 310).

Despite these serious hazards–known to supervisors and workers during the week prior to the incident–the low risk classification of the job was not reevaluated, nor did management formulate a plan to control hazards.

3.3.2 Good Practice Guidelines for Maintenance Work

A hazard evaluation is a formal analytical tool used to identify and examine potential hazards connected with a process or activity (CCPS, 1992; p. 7). The evaluation assists management in process plants in

---

[24] Tosco Avon MSDS for 50 Unit Naphtha, MSDS 1001, CSB 9914-E3-023, p. 2. Tosco Avon Safety Order S-29, Benzene, July 1998; Attachment 2.
[25] Tosco Avon Safety Order S-29, Benzene, July 1998; pp. 3, 6-7.

controlling hazards and preventing incidents. The Center for Chemical Process Safety (CCPS) describes evaluation techniques for identifying hazards in maintenance activity in process plants (CCPS, 1992; pp. 11, 428-430). The guidelines suggest a number of questions to be used in performing a hazard evaluation of maintenance work.[26]

In its good practice guide for hazard evaluation, the Institution of Chemical Engineers of the United Kingdom states:

> It is advisable to cover aspects of maintenance operations (with a HAZOP study), including isolation, preparation and removal for maintenance since these often create hazards as well as operability problems (ICE, 2000; p. 8).[27]

A number of factors may necessitate a hazard review of maintenance activity, including:

- Hazardous activities, such as hot work or hot tap and repair work on a live flare line (API, 1995b; pp. 2-3).

- Circumstances where existing procedures cannot be followed or where there are no applicable procedures (HSE, 1997; pp. 17-18).

- Situations where safety preconditions cannot be met, such as controlling ignition sources where flammables may be present.

Good practice guidelines (HSE, 1997, p. 18; see also CCPS, 1992, pp. 428-430) recommend that a comprehensive hazard evaluation include assessment of:

- Specific hazards introduced by performance of the maintenance work.

- Potential problems in achieving adequate isolation, such as depressuring, draining, and purging.

- Additional precautions, such as more frequent monitoring of the isolation, improved supervision, or contingency plans.

- The feasibility of safely conducting work while the process unit is in operation or postponing the job.

---

[26] CCPS guidance recommends that hazard evaluation questions include: What hazards are introduced by the maintenance activity? Is it necessary to completely shut down the process to safely conduct the repair?

[27] HAZOP (hazard and operability) is a well-recognized hazard evaluation technique.

3.3.3 British Petroleum Grangemouth Incident

An incident that occurred in 1987 at British Petroleum's Grangemouth refinery in Scotland was similar to the 1999 Tosco fire.[28] The job that led to the Grangemouth incident involved the attempted isolation of a section of a live flare line to remove a faulty valve. Four workers were killed after opening piping thought to be isolated and depressured. Although the isolation valves were placed in the closed position, plugging and valve leakage prevented complete isolation of the piping. The investigation report of the Health and Safety Executive of the United Kingdom made the following recommendations for senior management (HSE, 1989; pp. 13-14):

- Conduct a detailed hazard analysis during job planning.

- Before delegating work, develop detailed job instructions specific to the particular isolation to ensure the effective draining of flammable liquids.

- Maintain rigorous control over possible ignition sources in the vicinity of maintenance work.

3.4 Unit Shutdown Decision Making

Because of insufficient job planning and hazard evaluation, Tosco Avon refinery management did not recognize that safe conduct of the naphtha piping repair required shutting down the process unit. Once supervisors and workers knew that the line could not be drained or isolated, the unit needed to be shut down to safely conduct the repair. CSB recognizes that the shutdown and startup of an oil refinery process unit can introduce its own risks; however, the safe conduct of maintenance work requires a unit shutdown when serious hazards cannot otherwise be controlled or when the work cannot be deferred.

Because of insufficient job planning and hazard evaluation, Tosco Avon refinery management did not recognize that safe conduct of the naphtha piping repair required shutting down the process unit.

---

[28] See also "Lessons Learned From an On-Plot Refinery Tank Explosion," CCPS, 2000; p. 3. A lesson learned from the incident: "Pre-job hazard assessment should be conducted. Removing the strainer was a nonroutine task. The job planning and control did not include a discussion of the hazards by the personnel doing the job."

In spite of evidence that the line contained naphtha and was severely plugged, operations supervisors scheduled the piping removal. Multiple sources of ignition were as close as 3 feet from the repair work. Hot equipment surfaces, the most likely source of ignition, could be eliminated as a hazard only if the equipment was cooled, which required shutting down the unit. The work could not be deferred because the piping required immediate replacement. In an effective maintenance work evaluation process, CCPS recommends that management carefully consider whether it is "necessary to shut down the process completely to safely repair a piece of equipment" (CCPS, 1992; p. 429).

3.5 Management Oversight

A management system of accountability should include methods for establishing responsibility, evaluating performance, establishing feedback systems, and auditing.

3.5.1 Accountability for Hazardous Work

Despite significant hazards, Tosco management planned and executed the naphtha piping repair work as low risk maintenance. Under Tosco procedures, the unit operator was solely responsible for authorizing this work. Operations supervisors were minimally involved in planning or overseeing the line repair. No senior management or specialist personnel participated in assessing hazards. Although inspection personnel were included in decision making concerning the scope of the repair, their participation was limited to reviewing inspection data and determining what sections of the piping required immediate replacement.

Management oversight and accountability are essential elements of an effective PSM program. A management system of accountability should include methods for establishing responsibility, evaluating performance, establishing feedback systems, and auditing (CCPS, 1995c; p. 15).

At CSB's request, API and NPRA prepared a document on oil refining industry practices for authorizing repair and maintenance work. It states that for situations involving higher risk, such as hot work or the inability to isolate a leaking line, a higher degree of management scrutiny and approval may be needed: [29]

---

[29] See also CCPS, 1995b; p. 229.

Depending on the degree of risk some jobs may require, at a minimum, the approval of both a senior level safety person and the operations manager to deviate from routine or defined work practices. Higher risk jobs may require a risk management team including both labor and management level persons and safety, operations, maintenance, engineering, metallurgy and other disciplines depending on the nature of the work request (API/NPRA, 2000).

If a multidisciplinary team and senior management had participated in evaluating hazards and determining whether to shut down the process unit to safely conduct the repair, it is likely that the job would not have been allowed to proceed and the incident would not have occurred.

3.5.2   Supervision of Job Execution

The conduct of hazardous nonroutine maintenance requires close supervision, including frequent monitoring and unscheduled checks (CCPS, 1995b, p. 212; Townsend, 1998, p. 49). At the Avon refinery, operations supervisors stated that they oversaw activities in the process units when requested by the operators or as needed. On the morning of the incident, operations supervisors did not oversee the naphtha piping removal. The operations supervisor responsible for coordinating maintenance was not at work on February 23; the person responsible for filling in during such absences did not visit the unit prior to the fire.

The shift supervisor phoned the No. 1 operator prior to initiation of the piping repair work, asking if there were any problems in the unit. The operator responded that there were none. The shift supervisor visited the unit the morning of the incident, but did not observe the piping repair activities, review the permit, or inquire about the status of the draining attempts that had been ongoing for over a week.[30] Also, no health and safety personnel visited the job site before the incident occurred. Despite the presence of a crane, a vacuum truck, and

---

[30] The job description of the shift supervisor states the he or she is "accountable for everything that takes place with their crew," including ensuring "that all equipment is prepared properly and timely, that permits are completed and signed as appropriate per scheduled maintenance plan."

numerous contractors, the maintenance supervisor was the only management representative present during the conduct of the repair work on February 23.

A number of other deviations from Tosco procedures and good practice guidelines occurred during the naphtha piping repair activities. Although the following deviations were not directly causal to the incident, they demonstrated a pattern of serious management oversight deficiencies regarding maintenance activities at the Avon refinery:

- Naphtha was not recognized as a benzene stream and a health hazard. None of the piping repair work activities adhered to Tosco's benzene procedure, which required a special hazard permit and safety precautions, such as engineering controls, a benzene regulated area, personal protective equipment (PPE), and benzene exposure monitoring. [31]

- The use of a vacuum truck on February 23 was not included in the work permit, nor was a mobile entry permit issued, as required by Tosco procedures. [32]

- The use of a pneumatic saw in the piping removal was not included in the work permit, another requirement under Tosco procedures.[33]

- Several special hazard work permits authorized for the naphtha piping repair were not signed by the shift supervisor, as required by Tosco procedures.

Tosco procedures delegated to the operator the primary responsibility for identifying and controlling the hazards present in hazardous nonroutine repair. Process safety expert Frank Lees advises:

> It is desirable to include a caution to the effect that . . . although work may be delegated, responsibility remains with him (supervision); an indication of the levels of hazard so that high hazard situations are highlighted and those involved are prompted to consider whether there are other parties who should be consulted (Lees, 1996; p. 21/16).

---

[31] Tosco Avon Safety Order S-29, Benzene, July 1998.

[32] Tosco Avon Procedure TRFE003, Procedure for 100-Barrel Vacuum Truck, September 1995.

[33] Tosco Avon Procedure PFFE005, Procedure for Portable Power Pneumatic Hacksaw, December 1995.

Inadequate supervision was one of the issues the U.S. Environmental Protection Agency (EPA) investigated in its analysis of a fatal incident in the hydrocracker unit at the Avon refinery in 1997. EPA reported that inadequate supervision was a causal factor in the failure of a reactor effluent pipe and one fatality (USEPA, 1998; p. 65). The report stated that supervisor oversight of operations was deficient and contributed to the lack of adherence to emergency procedures. EPA concluded that there was no supervision at the unit during the hazardous abnormal situation, even though there had been a succession of operating problems prior to the temperature excursion that led to the pipe failure and fire.

3.5.3　Stop Work Authority

Tosco workers involved in the piping repair stated that they felt pressure to promptly execute the job because the unit was the only crude unit operating at that time.[34] Pressure to complete the job was also created by the presence of the vacuum truck, a crane, and contract workers.

Tosco management stated that workers had the authority to stop unsafe work activity and should have stopped the line replacement job. However, stop work authority–though a desirable safety policy if properly encouraged–is a less effective measure for incident prevention than good job preplanning for the following reasons (HSE, 1985, p. 11; CCPS, 1995c, p. 17):

> *Stop work authority–though a desirable safety policy if properly encouraged–is a less effective measure for incident prevention than good job preplanning.*

- It is exercised during the execution of work, when pressures to get the job done are generally greater.[35]

- It relies on the assertiveness of individual workers. To attempt to stop a job, a worker may need to assert a position that runs contrary to direct instructions from a supervisor.

- Once the job has begun, the idling of contractors and equipment can result in significant financial cost to the facility, which can add to the pressure to get the job done without delay.

---

[34] The No.3 unit was shut down in December 1998 and was being decommissioned.
[35] In discussing the management dilemma of production versus process safety, CCPS guidelines state: "The continuity of operations can best be addressed at the planning stage." (CCPS, 1995c; p. 17)

3.5.4 Auditing

The Avon refinery's safety auditing program consisted of undocumented observations (referred to as "layered safety surveys"). [36] These observations focused on worker behavior rather than measuring the effectiveness of procedures; they did not record findings, make recommendations, or track corrective actions.

In 1995, Tosco conducted a documented audit of its PSM program, as required every 3 years by Cal/OSHA. Tosco did not conduct a PSM compliance audit in 1998. Furthermore, neither Tosco Corporation nor Avon refinery management conducted documented audits of the facility's line breaking, lock-out/tag-out, or blinding procedures and practices in the 3 years prior to the incident.

Tosco's auditing program did not record or remedy the pattern of serious deviations from the safe performance of maintenance work and proper review of operational changes in process units. These deviations included:

- Opening of piping containing flammable liquids prior to draining.

- Transfer of flammable liquids to open containers proximate to sources of ignition.

- Inconsistent use of blind lists.

- Lack of supervisory oversight of hazardous work activities.

- Inconsistent use of MOC reviews for process changes.

Industry good practice guidelines recommend that an audit program consist of documenting findings, formulating corrective action to improve performance, and instituting followup controls.

Safety audits are an essential feedback mechanism for the effective functioning of a facility's safety management system. Industry good practice guidelines recommend that an audit program consist of documenting findings, formulating corrective action to improve performance, and instituting followup controls (CCPS, 1995c; pp. 313, 316).

Effective audits would have likely detected the inconsistent adherence to procedures at the Avon refinery and could have corrected these problems prior to the incident.

---

[36] Tosco asserted that some other documented audits were conducted at the Avon refinery, but did not provide CSB with evidence of such audits. Interview evidence indicates that no audits were conducted other than the layered safety surveys and the 1995 compliance audit.

3.6.1   Permit and Procedure Deficiencies

Deficiencies in two key elements of the maintenance work system—permitting and line breaking—contributed to causing the refinery fire.

The Avon refinery used written procedures, including a permitting system, to prepare equipment for maintenance work. The safe work procedure, which governed the work permitting system, applied to "all low risk and special hazard work."[37] The opening of pipelines or equipment required permitting. The procedure stated that the operator must ensure that equipment is depressured, drained, flushed, and purged of chemicals as completely as possible.

The authority to approve and issue work permits was generally delegated to operators; however, some types of higher hazard work required the approval of the shift supervisor or a health and safety department specialist. Higher hazard work included jobs that required entry into confined spaces, jobs that involved high energy hot work,[38] and those categorized as special hazard (e.g., opening live flare lines, radiography, and exposure to toxic materials such as lead, asbestos, benzene, and butadiene).

Deficiencies in the permitting system at the Avon refinery were exemplified by the permit issued on the day of the incident, which listed three tasks with different preparation requirements. By listing draining and removal together, the permit allowed both activities to be authorized even though draining was required before removal of the piping.[39]

The following deviations from good practice occurred with regard to line breaking and contributed to causing the incident:

- Hazardous nonroutine maintenance work was executed without a review of the job or permit authorization by an operations supervisor.

---

[37] Tosco Avon Procedure S-5, Safety Orders, Departmental Safe Work Permits, October 1998.

[38] Jobs that might result in sparks are defined as low energy hot work. Tasks involving the use of direct flames (such as torch cutting or welding) are defined as high energy hot work.

[39] Tosco procedures did not restrict work authorizations to one job per permit.

- Neither Tosco procedures nor the permit clearly indicated that personnel were to eliminate or effectively control sources of ignition prior to opening equipment.

- No course of action was specified if the required preparatory steps for opening piping, such as draining, could not be accomplished.

3.6.1.1 Inability to Follow Procedures

Good practice guidelines on equipment opening recommend that permits and procedures provide direction as to what course of action to take if existing hazards cannot be controlled or new hazards arise (Lees, 1996; p. 21/22). If the hazards require variation from the normal level of isolation, the work should be stopped and a hazard evaluation conducted by an appropriate level within management (HSE, 1997; p. 17). Tosco's permit system and procedures did not provide direction on what course of action to take if a line could not be drained.

3.6.1.2 Identification of Specific Hazards

In addition, industry good practice guidelines recommend that permits and procedures identify the specific hazards that may be encountered (Lees, 1996; p. 21/22). Neither Tosco procedures nor the permit form addressed the hazards of open containers of flammables or ignition sources from hot equipment surfaces, which were as close as 3 feet from the piping removal work. Workers were directed by the maintenance supervisor to drain the highly flammable liquid into an open plastic pan with multiple sources of ignition nearby. Process safety expert Trevor Kletz notes the often-unrecognized hazards of open containers (Kletz, 2000; p. 4). He emphasizes that open containers of flammable liquids should not be used in process plants because of the many potential ignition sources.

Another potentially hazardous activity was the transfer of naphtha to the vacuum truck, which was parked approximately 20 feet from the fractionator. Tosco procedures did not contain spacing requirements for placement of the truck. Good practice guidelines recommend that vehicles used for transferring flammable liquids not be allowed within at least 100 feet of sources of ignition (API, 1999; p. 9).

The potential hazard of static electricity was another issue not addressed by procedures or the permit system. Just prior to the incident, a plastic pan and sheeting were used to drain naphtha from the flange located near the fractionator. Transferring flammable liquids to a container such as a plastic pan or the use of plastic sheeting–both of which have insulating properties–may generate a static electrical charge. Furthermore, splashing of the liquid may also generate static electricity (NTSB, 1999; p. 2).

3.6.2 Deviations From Good Practice

Failure to drain the line prior to opening was another deviation from good practice. On several occasions during the course of the repair work, equipment had been opened prior to draining. On February 19, a small section of piping was removed before draining in an unsuccessful attempt to unplug the line. On the morning of the incident, the maintenance supervisor directed workers to cut and remove the top 9-foot section of the naphtha piping.[40] Personnel working on the removal job were aware that the piping contained naphtha. Two flanges were opened in an attempt to remove the naphtha, and another flange was opened when the top section of piping was removed. Tosco procedures required draining prior to opening equipment or using a pneumatic saw. [41]

It was a historical practice at the Tosco refinery to sometimes open equipment containing flammable liquids prior to draining. When drain lines were plugged or not available, witnesses described open-ing flanges in operating process units to release flammable liquids into an open container or onto the ground. Supervisors and workers did not perceive that this departure from Tosco procedures was a serious hazard.

---

[40] The pipe was cut using a pneumatic saw.

[41] Tosco Avon Procedure S-5, Safety Orders, Departmental Safe Work Permits, October 19, 1998; and PFFE005, Procedure for Portable Power Pneumatic Hacksaw, December 1995.

Good practice guidelines for process plants recommend that flanges not be opened or lines cut prior to draining flammable liquids (Lees, 1996; p. 21/26). Moreover, draining of flammables should take place through a closed system so as to shield the liquid from sources of ignition (Amoco, 1984; p. 13).[42] In addition, the use of a flange to drain flammable liquids in a running process unit with nearby sources of ignition is an unsafe practice because neither the rate nor the direction of flow can be adequately controlled, and it may be difficult to quickly stop the flow if needed.

3.7 Corrosion Control and Mechanical Integrity

The accelerated rate of corrosion in the naphtha piping was predominantly caused by a decrease in desalter performance and the entry of excessive amounts of water and corrosives into the fractionator.

3.7.1 Desalter Performance

A desalter is a crude oil processing vessel that reduces corrosion, plugging, and fouling of piping and equipment by removing inorganic salts, water, suspended solids, and water-soluble trace metals. The accelerated rate of corrosion in the naphtha piping was predominantly caused by a decrease in desalter performance and the entry of excessive amounts of water and corrosives into the fractionator (Hendrix, 2000; p. 1).

In the year prior to the incident, the desalter was run 40 percent beyond design capacity using heavier crude oils. The API gravity of the crude feed to the unit dropped on average from 27.2° in 1997 to 23.7° in 1998. Heavier oils with a lower API gravity are more difficult to separate from water, which impedes the desalting process.

Two internal incident reports describe desalter upsets that were directly related to crude feed and vessel problems at the Avon refinery. A September 1998 report recommended better dewatering of the crude. A March 1998 report described a serious incident when the gravity of the feed to the unit fell to 18° API. The report stated

---

[42] As Amoco reports: "Some equipment drains used during the shut down operation may not have permanent connections to a pump-out or closed drain system. If the material released from these drains can burn and then injure persons and damage equipment, install temporary facilities to drain the material to a closed system or another safe place."

that increased corrosion rates could be expected, specifically in the fractionator overhead and naphtha systems.

Operating logs for the unit noted more than two dozen desalter upsets during 1998. Performance deteriorated severely late in the year when the No. 3 crude unit was shut down. A process engineer described the desalter performance as "hopeful to non-existent." The chemical contractor for the desalter, Nalco/Exxon, also documented concerns about performance in a November 1998 memo, which stated that efforts to run the desalters efficiently had "never been more difficult."[43]

In a memo written in November 1998, Tosco management identified several potential improvements for immediate study and evaluation. These included operating the desalter vessels in parallel instead of in series, relocating a desalter from the No. 3 unit, and changing the electrolytic technology. Proposed solutions related to ongoing dewatering problems were also identified. Although Tosco management recognized the operational problems with the desalter, they did not adjust their equipment inspections accordingly;[44] nor did they implement corrective actions in a timely way to prevent material from plugging the pipe and to prohibit excessive corrosion in the unit.

A process engineer described the desalter performance as "hopeful to non-existent."

3.7.2 Corrosion

Maintenance records and notations in the operator's logbook revealed that as early as May 1998 the naphtha stripper level control valve (valve D) did not allow sufficient flow to maintain a liquid level inside the naphtha stripper. The bypass was run in the partially open position for at least 10 months prior to the incident, and the valve became plugged with solid corrosion deposits. The piping near valve D and the associated drain valves eventually became totally plugged. Long-term use of the partially open bypass valve also made it susceptible to erosion/corrosion.

---

[43] Nalco/Exxon Energy Chemicals memorandum to Tosco Corporation, "50 Unit Desalters Status," November 10, 1998.

[44] CSB investigators retained The Hendrix Group to examine corrosion and mechanical integrity issues related to this incident. The Hendrix Group found shortcomings (Appendix C) with the unit inspection program. However, CSB concluded that these problems were not directly causal to the fire.

Following the incident, Cal/OSHA commissioned a metallurgical analysis of the failed piping and components. It was determined that the bypass valve was eroded to such an extent that–when closed–it leaked (Figure 7) at a rate equivalent to a 1.5-inch-diameter hole (FTI Anamet, 1999; p. i).

It was determined that the bypass valve was eroded to such an extent that–when closed–it leaked at a rate equivalent to a 1.5-inch-diameter hole.

Figure 7. Leak test of the naphtha stripper level control bypass valve (B) in the closed position, showing significant water flow.
The inset photo highlights the gap between the seat and disc of the bypass valve in the closed position. This gap was equivalent to a 1.5-inch-diameter hole.

FTI Anamet determined that excessive amounts of ammonium chloride in the naphtha intensified the corrosive activity. Multiple analyses of residue specimens from the line were found to have very high chloride contents. This corrosive salt found its way into the fractionator and naphtha draw piping when the overhead reflux contained excessive water due to a large volume of water in the crude feed. The combination of corrosive salts and water in the naphtha piping led to excessive accelerated oxidation, which produced the original leak as well as the plugging in the piping and erosion/corrosion in the bypass valve. CSB investigators determined that the naphtha line was plugged with iron oxide, ammonium chloride, and sulfur compounds, which were either corrosive materials or products of corrosion.

In recognizing problems with chloride salt accumulation and plugging in the naphtha section of the fractionator tower, Tosco Avon management developed a water washing procedure to flush chlorides from the naphtha section of the tower.[45]

The combination of corrosive salts and water in the naphtha piping led to excessive accelerated oxidation, which produced the original leak as well as the plugging in the piping and erosion/corrosion in the bypass valve.

3.8 Management of Change

Tosco Avon management did not conduct an MOC review of the potential safety effects on the fractionator and associated piping that might result from:

- Operating the desalter beyond its design parameters.
- Increasing water in the crude feed.
- Shutting down the No. 3 unit and resulting effects on the 50 Unit.

API Recommended Practice 750 recommends that refiners review hazards that may be introduced as a result of projects or changes in operating conditions that increase throughput or accommodate different feedstocks (API, 1990; p. 4).

The Avon refinery's MOC program required an MOC review to be performed with a change in feedstocks.[46] Moreover, Tosco's program and API 750 stated that an MOC review should occur prior to changing design conditions. A Nalco/Exxon memo in December 1998 stated that the crude feed to the desalters was further increased to 55 to 80 percent over design specifications.[47] Not conducting an

---

[45] Tosco Avon Procedure 16-MS-06, Water Washing the Main Fractionator, September 1998.

[46] Tosco Avon Safety Order S-12, Management of Change Policy, March 1998; p. 8.

[47] Nalco/Exxon Energy Chemicals memorandum to Tosco Corporation, "50 Unit Desalters Report," December 7, 1998.

MOC review of changes in the feedstocks contributed to causing excessive rates of corrosion in the naphtha piping.

In addition, management did not conduct an MOC review for the process change of running with the naphtha stripper level control bypass valve partially open for a prolonged period. API 750 recommends conducting an MOC review for changes in technology that include "bypass connections around equipment that is normally in service" (API, 1990; p. 5). Not conducting an MOC review for operation of the bypass valve in the partially open position for months at a time resulted in the buildup of semisolid material in the control valve piping and drain lines, as well as erosion/corrosion of the valve seat and disc.

API 750 recommends conducting an MOC review for changes in technology that include "bypass connections around equipment that is normally in service."

4.0   Root and Contributing Causes

1. Tosco Avon refinery's maintenance management system did not recognize or control serious hazards posed by performing nonroutine repair work while the crude processing unit remained in operation.

    ■ Tosco Avon management did not recognize the hazards presented by sources of ignition, valve leakage, line plugging, and inability to drain the naphtha piping. Management did not conduct a hazard evaluation of the piping repair during the job planning stage. This allowed the execution of the job without proper control of hazards.

    ■ Management did not have a planning and authorization process to ensure that the job received appropriate management and safety personnel review and approval. The involvement of a multidisciplinary team in job planning and execution, along with the participation of higher level management, would have likely ensured that the process unit was shut down to safely make repairs once it was known that the naphtha piping could not be drained or isolated.

    ■ Tosco did not ensure that supervisory and safety personnel maintained a sufficient presence in the unit during the execution of this job. Tosco's reliance on individual workers to detect and stop unsafe work was an ineffective substitute for management oversight of hazardous work activities.

    ■ Tosco's procedures and work permit program did not require that sources of ignition be controlled prior to opening equipment that might contain flammables, nor did they specify what actions should be taken when safety requirements such as draining could not be accomplished.

2. Tosco's safety management oversight system did not detect or correct serious deficiencies in the execution of maintenance and review of process changes at its Avon refinery.

    Neither the parent Tosco Corporation nor the Avon facility management audited the refinery's line breaking, lockout/tagout, or blinding procedures in the 3 years prior to the incident. Periodic audits would have likely detected and corrected the pattern of serious deviations from safe work practices governing repair work and operational changes in process units. These deviations included practices such as:

- Opening of piping containing flammable liquids prior to draining.

- Transfer of flammable liquids to open containers.

- Inconsistent use of blind lists.

- Lack of supervisory oversight of hazardous work activities.

- Inconsistent use of MOC reviews for process changes.

4.2 Contributing Causes

1. **Tosco Avon refinery management did not conduct an MOC review of operational changes that led to excessive corrosion rates in the naphtha piping.**

   Management did not consider the safety implications of process changes, such as:

   - Running the crude desalter beyond its design parameters.

   - Excessive water in the crude feed.

   - Prolonged operation of the naphtha stripper level control bypass valve in the partially open position.

   These changes led to excessive corrosion rates in the naphtha piping and bypass valve, which prevented isolation and draining of the naphtha pipe.

2. **The crude unit corrosion control program was inadequate.**

   Although Avon refinery management was aware that operational problems would increase corrosion rates in the naphtha line, they did not take timely corrective actions to prevent plugging and excessive corrosion in the piping.

5.0   Recommendations

Conduct periodic safety audits of your oil refinery facilities in light of the findings of this report. (1999-014-I-CA-R1) At a minimum, ensure that:

- Audits assess the following:

  ▲ Safe conduct of hazardous nonroutine maintenance

  ▲ Management oversight and accountability for safety

  ▲ Management of change program

  ▲ Corrosion control program.

- Audits are documented in a written report that contains findings and recommendations and is shared with the workforce at the facility.

- Audit recommendations are tracked and implemented.

1.   Implement a program to ensure the safe conduct of hazardous nonroutine maintenance. (1999-014-I-CA-R2) At a minimum, require that:

- A written hazard evaluation is performed by a multi-disciplinary team and, where feasible, conducted during the job planning process prior to the day of job execution.

- Work authorizations for jobs with higher levels of hazards receive higher levels of management review, approval, and oversight.

- A written decision-making protocol is used to determine when it is necessary to shut down a process unit to safely conduct repairs.

- Management and safety personnel are present at the job site at a frequency sufficient to ensure the safe conduct of work.

- Procedures and permits identify the specific hazards present and specify a course of action to be taken if safety require-ments—such as controlling ignition sources, draining flam-mables, and verifying isolation—are not met.

- The program is periodically audited, generates written findings and recommendations, and implements corrective actions.

2. Ensure that MOC reviews are conducted for changes in operating conditions, such as altering feedstock composition, increasing process unit throughput, or prolonged diversion of process flow through manual bypass valves. (1999-014-I-CA-R3)

3. Ensure that your corrosion management program effectively controls corrosion rates prior to the loss of containment or plugging of process equipment, which may affect safety. (1999-014-I-CA-R4)

American Petroleum Institute (API)

Paper, Allied-Industrial, Chemical & Energy Workers
  International Union (PACE)

National Petrochemical & Refiners Association (NPRA)

Communicate the findings of this report to your membership. (1999-014-I-CA-R5)

By the

U.S. CHEMICAL SAFETY AND HAZARD INVESTIGATION BOARD

Gerald V. Poje, Ph.D.
Member

Isadore Rosenthal, Ph.D.
Member

Andrea Kidd Taylor, Dr. P.H.
Member

March 21, 2001

6.0 References

American Petroleum Institute (API) and National Petrochemical & Refiners Association (NPRA), 2000. *Work Authorization in Refineries,* prepared for U.S. Chemical Safety and Hazard Investigation Board (CSB), July 2000.

American Petroleum Institute (API), 1999. *Safe Operation of Vacuum Trucks in Petroleum Service,* API Publication No. 2219.

American Petroleum Institute (API), 1995a. *Procedures for Welding or Hot Tapping on Equipment in Service,* API Recommended Practice 2201, Fourth Edition, September 1995.

American Petroleum Institute (API), 1995b. *Safe Welding, Cutting, and Other Hot Work Practices in Refineries, Gas Plants, and Petrochemical Plants,* API Publication No. 2009, Sixth Edition.

American Petroleum Institute (API), 1990. *Management of Process Hazards,* Recommended Practice 750, First Edition, 4.2.2(b).

Amoco Oil Company, 1984. *Safe Ups and Downs for Refinery Units.*

Center for Chemical Process Safety (CCPS), 2000. "Lessons Learned From an On-Plot Refinery Tank Explosion," by K. Ann Paine, *Process Industry Incidents,* American Institute of Chemical Engineers (AIChE).

Center for Chemical Process Safety (CCPS), 1995a. *Guidelines for Process Safety Documentation,* American Institute of Chemical Engineers (AIChE).

Center for Chemical Process Safety (CCPS), 1995b. *Guidelines for Safe Process Operations and Maintenance,* American Institute of Chemical Engineers (AIChE).

Center for Chemical Process Safety (CCPS), 1995c. *Plant Guidelines for Technical Management of Chemical Process Safety,* American Institute of Chemical Engineers (AIChE).

Center for Chemical Process Safety (CCPS), 1992. *Guidelines for Hazard Evaluation Procedures,* American Institute of Chemical Engineers (AIChE).

Contra Costa County, California, 1999. *Investigation Into the Causes of the Fire of February 23, 1999, at No. 50 Crude Unit, Tosco Avon Refinery,* Contra Costa Health Services, July 1999.

FTI Anamet, 1999. *Metallurgical Evaluation of Naphtha Draw Line/ Valve and Analyses of Petroleum Samples From a Crude Unit* at the Tosco Avon Oil Refinery, prepared for California Department of Industrial Relations, Division of Occupational Safety and Health.

Health and Safety Executive (HSE), 1997. *The Safe Isolation of Plants and Equipment, Oil Industry Advisory Committee,* Norwich, U.K.: HSE Books.

Health and Safety Executive (HSE), 1989. *The Fires and Explosion at BP Oil (Grangemouth) Refinery Ltd.,* Norwich, U.K.: HSE Books.

Health and Safety Executive (HSE), 1985. *Deadly Maintenance,* Norwich, U.K.: HSE Books.

Institution of Chemical Engineers (ICE), 2000. *HAZOP: Guide to Best Practice, Guidelines to Best Practice for the Process and Chemical Industries,* Rugby, U.K.

Kletz, Trevor A., 1989. *What Went Wrong? Case Histories of Process Plant Disasters,* Second Edition, Gulf Publishing Company.

Lees, Frank P., 1996. *Loss Prevention in the Process Industries: Hazard Identification Assessment and Control,* Vol. 2, Second Edition, Oxford, U.K.: Butterworth-Heinemann.

National Transportation Safety Board (NTSB), 1999. *Safety Recommendation H-99-42,* to the International Association of Fire Chiefs, October 1, 1999.

Occupational Safety and Health Administration (OSHA), 1990. *The Phillips 66 Company Houston Chemical Complex Explosion and Fire, A Report to the President,* U.S. Department of Labor.

The Hendrix Group, Inc., 2000. *Hendrix Report to the CSB.*

Townsend, Arthur, 1998. *Maintenance of Process Plant,* Second Edition, Rugby, U.K.: Institution of Chemical Engineers (ICE).

U.S. Environmental Protection Agency (USEPA), 1998. *Chemical Accident Investigation Report, Tosco Avon Refinery, Martinez, California,* November 1998.

1. **February 10, 1999, Wednesday**

    a. 1:20 pm: A leak was detected in the 50 Unit at the first elbow of the naphtha piping leaving the crude fractionator tower (just downstream of valve A; Figure 2).

    b. Emergency responders arrived at the scene of the leak with firefighter personal protective equipment (PPE) and self-contained breathing apparatus (SCBA). Fire hoses and a snorkel truck were set up in case of a fire and used to wash down the fractionator tower and decks.

    c. The following valves were placed in the closed position to isolate the naphtha piping–the block valve on the naphtha draw line near the fractionator (valve A), block valves C and E upstream and downstream of the naphtha stripper level control valve (valve D or LCV-150), and the naphtha stripper level control bypass valve (valve B).[1] The naphtha piping appeared to stop leaking. No clamp was installed on the leaking section of the pipe.

    d. 2:25 pm: Work began to strip insulation from the naphtha piping. The operations superintendent and the superintendent of shift operations were on scene during isolation of the line and at the beginning of the insulation removal work.

    e. An emergency work order was requested to replace the naphtha piping.

    f. The naphtha piping was inspected using ultrasonic and radiographic testing to identify the extent of wall thinning.

    g. 9:40 pm: The liquid in the naphtha stripper vessel rose to a high level. Operations personnel lowered the liquid level by opening the naphtha to storage flow control valve (valve J; Figure 2), downstream of the naphtha stripper.

2. **February 11, Thursday**

    a. As a result of the initial inspection, a decision was made to replace all of the naphtha piping from the fractionator to the naphtha stripper.

---

[1] Valves A, B, C, and E (Figure 2) are also referred to as the isolation valves.

b. Contract workers began erecting scaffolding on the fractionator to provide access to the piping.

3. **February 12, Friday:** The liquid level in the naphtha stripper increased again and was lowered by operations personnel.

4. **February 13, Saturday**

   a. A No. 1 operator observed naphtha "misting" from the hole on the naphtha line at the site of the original leak on February 10. The naphtha piping felt warm to the touch. The No. 1 operator and the shift supervisor tightened the isolation valves with a wrench and an extension in an attempt to stop the leak. The leak appeared to subside.

   b. The operator logbook noted that "the ruptured draw line is full" in reference to the naphtha piping that had been leaking.

   c. The high naphtha stripper level was lowered after retightening of the isolation valves (see 4.a above). The shift supervisor's log (referred to as "area notes"), available electronically, recorded that the naphtha stripper level was lowered. The shift supervisor stated that the block valves isolating the naphtha piping might have been leaking.

5. **February 13 and 14:** During the night shift into the morning of February 14, the operators lowered the level in the naphtha stripper on four different occasions. After the fourth occurrence, the naphtha flow control valve (valve J) was left open so that the naphtha could flow through the pump to storage, thus preventing the stripper from refilling.

6. **February 16, Tuesday:** The No. 1 operator attempted to drain the naphtha piping from drain valves F and G on either side of the naphtha stripper level control valve. A hose was attached to the drain valves running to the ground level. No liquid was removed. The No. 1 operator informed the business team leader that the naphtha drain lines were plugged.

7. **February 16 and 17:** The job scope was reduced after it was determined that portions of naphtha piping could not be isolated to allow replacement of all the piping while the unit was running.

Tosco inspectors reevaluated the thickness data and concluded that the portion of piping between the naphtha stripper level control valve (valve D) and the naphtha stripper did not need immediate replacement.

8. February 17, Wednesday

    a. The maintenance supervisor observed a small stream of naphtha intermittently draining from the point of the original leak. The line felt warm to the touch, and the maintenance supervisor assumed that the block valve (valve A) on the naphtha piping near the fractionator was leaking from the fractionator. The operator logbook recorded that isolation valves were again retightened (valves A and B).

    b. The No. 1 operator opened the drain valves (valves F and G) on either side of the naphtha stripper level control valve (valve D). When no flow was observed, the operator used a welding rod[2] to attempt to clear the plugging in the drain lines. Again, no flow was observed. It was recorded in the logbook that the drain lines were plugged and could not be cleared.

    c. The failed attempt to drain the naphtha piping was communicated by two No. 1 operators to the operations supervisor and the maintenance supervisor. The operators presented a plan to shut down the unit if the plugging could not be cleared. The operations supervisor initiated a request for maintenance workers to clear the drain lines (connected to valves F and G).

    d. Maintenance personnel began to sketch and detail the specifications of the naphtha piping for replacement.

9. February 18, Thursday, noon: Maintenance workers were in the unit to "unplug 1-inch drain valves (valves F and G) and drain the 6-inch naphtha piping on the fractionation tower." After repeated unsuccessful attempts to drill out the plugged drain lines near the naphtha stripper level control valve (valve

---

[2] The use of a wire or rod to unplug a drain line is an unsafe procedure (Amoco, 1984; p. 49).

D), the reaming device broke. The safe work permit was marked as "job not finished."

10. **February 19, Friday**

   a. In the maintenance work schedule report for the following week, the maintenance lead planner requested a crane to remove naphtha piping for Tuesday, February 23.

   b. 12:05 pm: In response to unsuccessful attempts to unplug the drain lines, a safe work permit was issued to remove a short piping spool piece downstream of the naphtha stripper level control valve (between valves D and E). This work was directed and witnessed by the maintenance supervisor, who signed into the unit for 2 hours to oversee the work.

   c. The spool piece (between valves D and E) was removed. The block valves (valves C and E) were not locked out. Block valve C was observed by the maintenance supervisor to be jammed partially open. The spool piece was not drained, nor was isolation of the block valves verified prior to removal. The spool piece was full of semisolid material, which plugged the line. A blind flange with a drain valve (valve I; Figure 4) was installed on the downstream side of the naphtha stripper level control valve (valve D). No attempts were made to drain the line after this activity. The safe work permit was marked "job not finished."

11. **February 22, Monday**

   a. The operations supervisor prepared a permit readiness sheet with input from the maintenance supervisor. The sheet stated, "Bigge, Interstate Scaffold, Tosco and Rust personnel to drain and start removal of naphtha draw piping." This document was available electronically and sent to the shift supervisor.

   b. The No. 1 operator observed the leak reoccur at the original location. The naphtha piping felt warm to the touch. The shift supervisor was brought up to the deck to observe the leak.

c. A hot work permit was issued to cut out a section of the deck on a platform on the fractionator tower, 107.5 feet above grade. To contain the naphtha while the cut was made, a plug was placed in the perforation of the piping where the leak had occurred. The maintenance supervisor directed the plug to be removed upon completion of the hot work.

d. The maintenance supervisor and the maintenance lead planner arranged for a vacuum truck from a contracting company for the next day.

e. An operator prepared a permit during the nightshift to "Erect scaffolding, drain and remove piping (naphtha draw)."

12. **February 23, Tuesday**

   a 7:20 am: A vacuum truck from Waste Management Industrial Services arrived at the unit.

   b. 7:40 am: Tosco maintenance employees arrived at the unit to "drain and remove naphtha piping."

   c. 8:00 am: A Bigge crane operator and rigger arrived at the unit to assist in removing the piping.

   d. 8:00 am: The operations process engineer visited the unit and discussed the naphtha piping replacement. An operator told him that several draining efforts had been unsuccessful and that the reaming device used to clear the drain lines (connected to valves F and G) had broken on February 18. The engineer suspected that the naphtha piping isolation valves were leaking. He was aware that naphtha was in the piping.

   e. 8:30 am: A maintenance worker and a No. 1 operator reviewed the job site and signed a safe work permit prior to the start of the job.

   f. 8:40 am: The maintenance supervisor entered the unit to supervise the naphtha piping replacement job.

   g. 8:50 am: A maintenance worker signed the work authorization permit.

   h. 9:19-9:26 am: Maintenance personnel initially attempted to remove naphtha from a drain valve (valve I; Figure 4) in the blind flange downstream of the naphtha stripper level control valve (valve D), where the spool piece was previously removed. No material was observed coming from the drain line (connected to valve I).

i. The workers attempted to wedge the flange open just upstream of the control valve (valve D). No material was observed coming from the flange.

j. 9:40 am: Before the line was drained or isolation was verified, maintenance workers, the maintenance supervisor, and the No. 1 operator ascended the tower to begin cutting the naphtha piping with a pneumatic saw. The maintenance supervisor showed the workers where to make the initial cut into the piping.

k. 9:50 am: The maintenance supervisor left the unit.

l. 10:15 am: The maintenance supervisor returned to the unit halfway through the first cut into the naphtha piping.

m. The first cut was completed at an approximate elevation of 104 feet above grade (Figure 6). The crane was used to remove the top 9-foot section of the piping.

n. The maintenance supervisor directed a second cut on the naphtha piping at an elevation of 79 feet above grade (Figure 6). The cutting was stopped when the blade pierced the pipe and a small amount of naphtha began to leak from the line.

o. The maintenance supervisor attempted to locate the liquid level in the line by tapping on the pipe with a hammer and listening to the change in sound. He believed that the naphtha level was just above the location of the second cut.

p. A third attempt was made to drain the piping at the location of the flange upstream of the naphtha stripper level control valve (valve D). No material was observed coming out of the flange. The maintenance supervisor and a mechanic attempted to use a scraping tool to unplug the line at the flange; however, the tool did not penetrate the hardened material plugging the piping.

q. 11:00-11:30 am: The maintenance crew broke for lunch, after which the maintenance supervisor discussed possible drain points and directed the workers to drain the piping from the flange closest to the fractionator (flange 2).

r. 11:45 am: The next attempt to drain was initiated at the base of the vertical run of piping close to the fractionator (flange 2), at an elevation of 38 feet above grade. Naphtha was drained into a plastic pan with the flow directed by plastic sheeting. The naphtha was suctioned from the pan with a hose connected to the vacuum truck, which was parked at ground level (Figure 4).

s. 12:18 pm: Naphtha started to flow very rapidly from the line at the open end of the pipe. Hot equipment surfaces most likely ignited the naphtha. The resulting fire engulfed workers on the fractionator tower, killing four men and seriously injuring another.

Tosco Fire, Avon Refinery

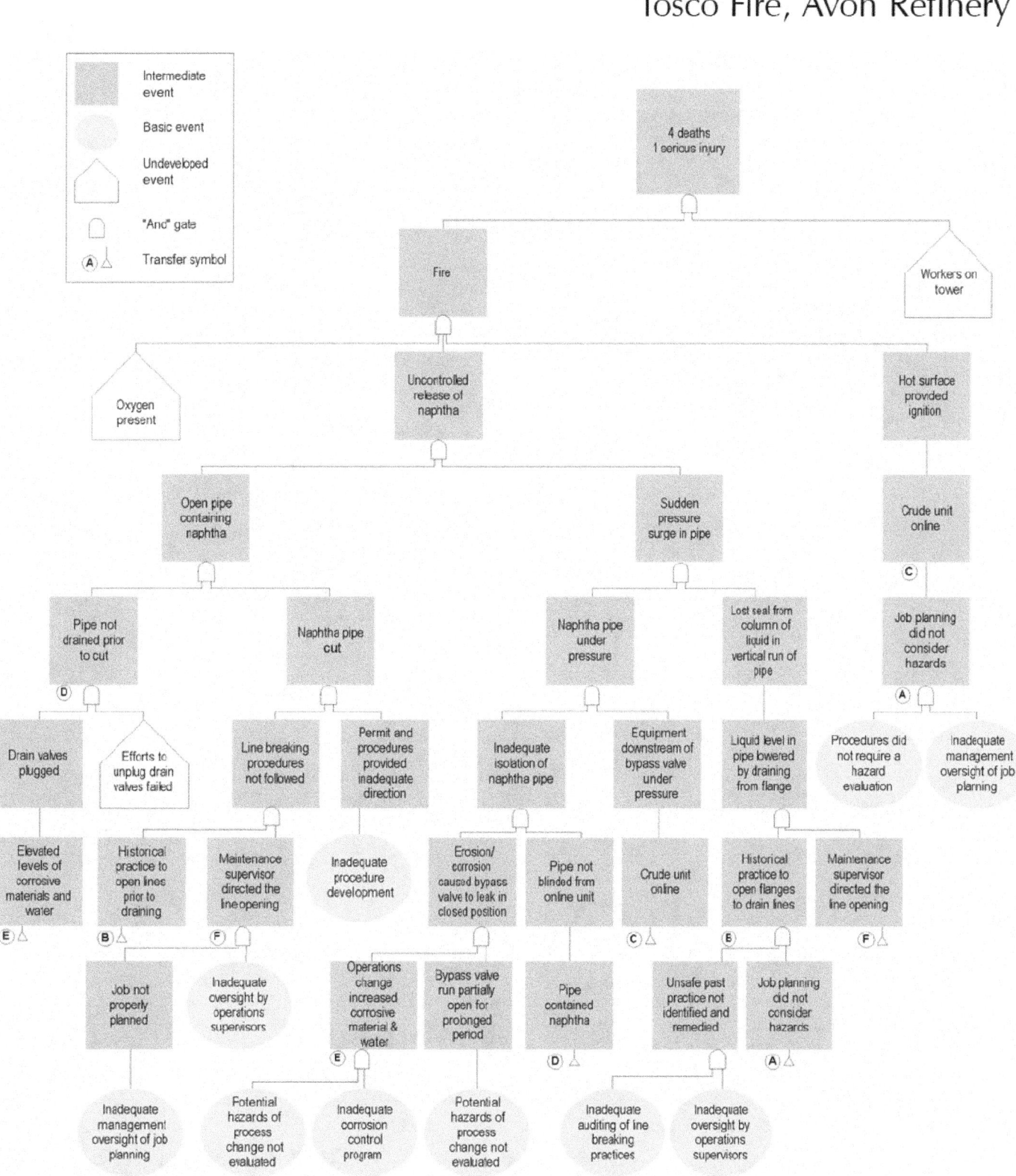

APPENDIX C: Executive Summary of The Hendrix Group, Inc., Report

This report documents the result of a technical review of documents associated with a naphtha leak and subsequent fire at the No. 50 Crude Unit at Tosco Refining Company's Avon Refinery located in Martinez California and the corrosion related and mechanical integrity issues that contributed to the fire. The results of the review showed that:

- The cause of the naphtha line leak precipitating replacement of the line was erosion-corrosion due to aqueous ammonium chlorides. The naphtha line leak, pluggage of the bleeder valves at the naphtha line control valve loop and erosion-corrosion of the bypass valve were all contributing causes leading to the incident. The control valve piping and bleeder valve pluggage and erosion-corrosion of the by-pass valve made the consequence of the incident greater, based on making draining more difficult and contributing to a greater amount of flammable liquid in the line than would otherwise be the case. The valve leak, the pluggage and the control valve erosion-corrosion were all due to the elevated levels of corrosive materials.

- Inadequate desalter operation with heavy crude slates directly contributed to the naphtha line corrosion by allowing excessive water and hydrolyzable chlorides to enter the fractionator tower, forming corrosive, acidic water in the top of the tower.

- Water slugs entering the tower, largely from inadequate dehydration of the crude feed by the desalters, caused tower upsets and water flooding of tower upper trays, resulting in water in sections of the tower where it normally would not be expected, including the naphtha draw line. However, there was significant available evidence to suggest the potential for corrosion of the naphtha draw line, including: (a) Tosco incident reports describing desalter problems with attending consequences of plant wide corrosion, (b) documented corrosion of fractionator tower trays in the vicinity of the naphtha draw line, (c) previous incidents of corrosion in the naphtha stripper and bottoms piping and, (d) having to drain water from the fractionator tower reflux line.

- Accelerated corrosion in the Main Fractionator and in associated overhead equipment had been a problem since the early

1980's. Tosco did not modify their corrosion control program to address continued equipment corrosion associated leaks.

- Tosco's mechanical integrity and inspection program failed to predict and locate corrosion problems before they resulted in leaks or emergency on-stream repairs. It was unclear from Tosco's inspection documentation what schedules were in place to conduct thickness surveys on the Naphtha Stripper draw line. Tosco had classified the line as a Class 1 line, with a maximum next inspection interval of 5 years, based on API 570, Piping Inspection Code. However, in their Piping Corrosion Management System (PCMS) documentation, (8/7/99) they appeared to list as much as a ten-year next inspection interval for the line. An inspection deficiency contributing to the incident, was the lack of a sufficient PCMS database at the time of the incident permitting corrosion rate determination.

- Failure of the corrosion control and corrosion monitoring programs to prevent events leading to the incident by practicing predictive inspection were symptomatic of: (a) inadequate management oversight, (b) inadequate or non existing documentation supporting SFAR-PSM-j, Mechanical Integrity, (c) insufficient inspection data documentation, (d) lack of proper inspection execution and, (e) inadequate communications between the mechanical integrity department and Unit 50 operations personnel.